1989

Fabricating
History

Fabricating History:
English Writers on the French Revolution

Barton R. Friedman

Princeton University Press
Princeton, New Jersey

Copyright © 1988 by Princeton University Press
Published by Princeton University Press,
41 William Street, Princeton, New Jersey 08540
In the United Kingdom: Princeton University Press,
Guildford, Surrey

This book has been composed in Trump Mediaeval

Clothbound editions
of Princeton University Press books
are printed on acid-free paper, and binding materials
are chosen for strength and durability

Printed in the United States of America
by Princeton University Press, Princeton, New Jersey

Library of Congress Cataloging-in-Publication Data
Friedman, Barton R.
Fabricating history : English writers on the French Revolution /
Barton R. Friedman.
p. cm.
Bibliography: p.
Includes index.
ISBN 0-691-06729-5 (alk. paper)
1. English literature—19th century—History and criticism.
2. France—History—Revolution, 1789–1799—Literature and the
revolution. 3. Napoleonic Wars, 1800–1814—Literature and the wars.
4. France in literature. 5. Revolutions in literature. 6. War in
literature. 7. Literature and history. 8. Historical fiction,
English—History and criticism. 9. English literature—French
influences. 10. France—History—Revolution, 1789–1799—
Historiography. 11. Napoleonic Wars, 1800–1814—Historiography.
I. Title.
PR129.F8F75 1988
820'.9'358—dc19   87-36123
CIP

IN
MEMORY
OF MY FATHER,
Abraham I. Friedman
1907–1981,

AND OF MY COUSIN,
Julian Friedman
1920–1983

Contents

Acknowledgments

I wish to thank the editors of *Clio*, who have kindly permitted me to reprint the chapter on *The Dynasts*, which originally appeared (in somewhat abbreviated form) as an article in that journal.

I also wish to thank my colleagues Louis T. Milic, who read and advised me on *The Dynasts* chapter at a time when it was refusing to come together; Steven Jeffrey Jones, who responded to my questions about literary theory as a good teacher should, by feeding me books to read; and especially Phillips Salman, who not only read the whole book in manuscript but listened patiently over endless cups of coffee while I talked and retalked ideas, themes, and critical arguments.

I am grateful as well to my secretary, Rita Hammond, who typed more drafts of this book than either of us cares to count.

But I am most grateful, as always, to my wife, Sheila, who tolerated me on those evenings when, as we sat at our dinner table, my mind was wandering among the Paris crowds or exploring some Napoleonic battlefield.

A Note on Texts

I have, for the reader's convenience, incorporated citations of the works central to this study into the text rather than including them in notes. They are abbreviated as follows:

Blake

B *William Blake's Writings*, ed. G. E. Bentley, Jr. Discussion of the illuminations in *America* and *Europe* refer to vol. 1, *Engraved and Etched Writings* (Oxford: Clarendon P, 1978).

E *The Poetry and Prose of William Blake*, ed. David V. Erdman. rev. ed. (Garden City, N.Y.: Doubleday, 1970). All textual references are to this edition.
 Am. *America*
 DC *A Descriptive Catalogue of Pictures*
 Eur. *Europe*
 FR *The French Revolution*
 MHH *The Marriage of Heaven and Hell*

Burke

RP *Letters on a Regicide Peace*; text from *The Writings and Speeches of Edmund Burke*, vols. 5 and 6 (Boston: Little, Brown, 1901).

RRF   *Reflections on the Revolution in France*, ed. William B. Todd (New York: Holt, Rinehart and Winston, 1959).

Carlyle

CME   *Critical and Miscellaneous Essays*, 5 vols., in *The Works of Thomas Carlyle*, centenary ed., vols. 26–30 (New York: AMS P, 1969).

Fr. Rev.   *The French Revolution*, 3 vols., in *Works*, centenary ed., vols. 2–4.

SR   *Sartor Resartus*, ed. Charles Frederick Harrold (New York: Odyssey P, 1937).

Coleridge

BL   *Biographia Literaria*, ed. James Engell and W. Jackson Bate; vol. 7 (in two parts) of *The Collected Works of Samuel Taylor Coleridge*, ed. Kathleen Coburn, assoc. ed. Bart Winer, Bollingen Series (Princeton: Princeton UP, 1970–1984).

SM   *The Statesman's Manual*, in *Lay Sermons*, ed. R. J. White, vol. 6 of *Collected Works*.

Paine

CS   *Common Sense* (Newport: Solomon Southwick, 1776).

RM   *The Rights of Man*, vol. 2 of *The Writings of Thomas Paine*, ed. Moncure Daniel Conway (New York: AMS P, 1967)

Citations of the works listed below are also incorporated into the text, but in contexts that identify them without the need to abbreviate titles:

Carlyle   *On Heroes, Hero-Worship, and the Heroic in History*, in *Works*, centenary ed., vol. 5.

Coleridge   *Lectures 1795 on Politics and Religion*, ed. Lewis Patton and Peter Mann, vol. 1 of *Collected Works*.

Dickens   *A Tale of Two Cities*, ed. George Woodcock (Harmondsworth, Middlesex: Penguin, 1970). Reprint of the first edition published in volume form (London: Chapman and Hall, 1859).

Hardy   *The Dynasts*, intro. John Wain (London: Macmillan, 1965).

Hazlitt   *The Complete Works of William Hazlitt*, ed. P. P. Howe, 21 vols. (New York: AMS P, 1967). *The Life of Napoleon Buonaparte*, vols. 13–15.

Marx   *The Eighteenth Brumaire of Louis Bonaparte* (New York: International Publishers, 1963).

Scott   *The Life of Napoleon Buonaparte*, 3 vols. (Philadelphia: Carey, Lea and Carey, 1827).

Citations of Blake's illuminated books (*America, Europe*, and *The Marriage of Heaven and Hell*) refer to plate and line number. Citations of his unengraved poem, *The French Revolution*, refer to manuscript page and line number, as specified in Erdman's edition. Citations of the unfinished play, *King Edward the Third*, refer to scene and line number, also as specified in Erdman's edition.

The volume numbers of Carlyle's *Works* indicate their place both in the sequence of the edition as a whole and, in the case of multivolume texts, in the sequence of the particular works to which they belong. Thus volumes 26–30 of *Works* are also volumes 1–5 of *Critical and Miscellaneous Essays*, and volumes 2–4 of *Works* are volumes 1–3 of *The French Revolution*. I cite these texts by volume number in the sequence of the work under scrutiny and by page number.

The system of reference used for the other works taken up in this study will be self-evident to the reader, with the exception perhaps of *The Dynasts*, which is cited by part, act, scene, and page number.

All italics are as in the original texts, unless otherwise specified.

Fabricating History

Introduction

Some years ago at a conference on history and narrative art, hosted by the University of Wisconsin English Department, I watched one of the few historians to have ventured into that den of (mostly) literary critics and philosophers of history berate the assemblage for repeatedly insisting on the kinship between historical narrative and fiction. Actually, the argument in which this indignant historian became embroiled had been adumbrated by Northrop Frye, who observes, in an essay entitled "New Directions from Old," that though historians' narratives (like those of poets) incorporate "unifying forms," or myths, "to tell a historian that what gives shape to his book is a myth would sound to him vaguely insulting."[1] My fellow conferee, convinced that, the only good history being scientific history, the integrity of his discipline was under siege, was more than just vaguely insulted: how could anyone claim a link between history and myth—fiction—when historical narratives consist of facts carefully documented?

Yet J. H. Hexter, himself (as he puts it) "a professional historian and a writer of professional history," claims exactly such a link, justifying for history the use of a rhetoric closer to the rhetoric of fiction than of science as "not only permissible but on occasion indispensable."[2] Hexter is not, of course, pronouncing historians liars. (Nor, I might add, were the targets of his colleague's ire in that Wisconsin lecture hall.) He is acknowledging that to capture historical truth, in its complexity and frequent ambiguity, often requires the techniques of art rather than of science.

Frye takes a complementary view. The writing of history, he insists in *Anatomy of Criticism*, remains an art, even though the evidence from which written history derives is gathered and scrutinized with the rigor of science. As Frye asserts in "New Directions from Old" (citing Menander's conundrum that, his new play having been finished, he had now only to write it):

> A historian in the position of Menander, ready to write his book, would say that he had finished his research and had only to put it into shape. He works toward his unifying form, as the poet works from it. The informing pattern of the historian's book, which is his *mythos* or plot, is secondary. Hence the first thing that strikes us about the relation of the poet to the historian is their opposition.[3]

Their opposition, however, is one of emphasis not of kind. Both Frye and Hexter are exploring the uncertain boundary between history and fiction. Both are indirectly asking, as over a century of philosophers (if seldom historians) before them had asked, what is history?

Embarking on my own exploration of that uncertain boundary, I soon discovered, as I suspect Frye and Hexter had, that to put the question so baldly was to put it badly—to invite simplistic, or at least overly restrictive, answers. David Hackett Fischer, in a book provocatively entitled *Historians' Fallacies*, also (and indirectly) asks, what is history? Answering that history comprises solutions to problems empirically resolvable, he concludes that historians can respond only to queries about what and how, never about why. Queries about why, he warns the incautious historian, lead to the trackless wood of metaphysics.[4]

Practitioners of the latest approach to charting the past, psychohistory, would no doubt dismiss Fischer's refusal to ask why as a kind of self-protective reflex. Peter Loewenberg reports finding that historians, scrupulous about determining whether an event has happened and when, often lapse into facile explanations of why it happened to repress uncomfortable or (they believe) irrelevant feelings about their subjects. The historian, he insists, must take into account his own innate biases as an element in the reality he discerns.[5]

No less distinguished a thinker about history than Henry Adams, though equipped with none of Loewenberg's psychoanalytic tools, would have agreed. Adams seeks to understand his civilization's present by understanding his own past. Recalling how as a young man he had rummaged through the debris of modern Rome to find the causes of ancient Rome's collapse, he insists that "the eternal question"—for Tacitus, Michelangelo, and Gibbon, as well as for himself—is always "Why! Why!! Why!!"[6]

Why is also recognized as an (if not the) eternal question by two contemporary thinkers about history, W. B. Gallie and Maurice Mandelbaum. They are concerned with the methods of selection and of construction by which historiography proceeds. A historian, Gallie argues, weighs the relative value of the data his research uncovers according to partly preconceived ideas of what will prove valuable. His ideas of what will prove valuable are, Mandelbaum proposes, partly determined by his knowledge of how things turned out: he starts, in effect, from the end and works back, reforging causal links as he goes.[7]

Hexter supplies a nice, if less than monumentally important, model of these historiographical practices in his treatment of an event he remembers with obvious fondness, the spectacular recovery of the New York Giants from a thirteen-game deficit to beat the Brooklyn Dodgers and win the 1951 National League pennant. Building from the climax of his drama (the moment in the ninth inning of the third game of a three-game playoff when Bobby Thomson hit his decisive home run off Ralph Branca), and reproducing the rising excitement of radio announcer Russ Hodges—"The Giants win the pennant! *The Giants win the pennant!* THE GIANTS WIN THE PENNANT!"— Hexter outlines the steps by which the baseball historian might plot his narrative of the Miracle of Coogan's Bluff.[8]

Hexter's historian, however, is writing not the history but only a history of the 1951 National League pennant race; and one wonders whether reading his book would differ from spending a few hours in a library turning the August and September sports pages of the *New York Times*. It may be I can show the limitations of his approach by fabricating a small historical fiction, though one not (I would contend) incompatible with historical truth. Since, as a friend who is a specialist in the Tory side of the American Revolution likes to remind me, history is usually written by the winners, we start with the postgame scene in the Giants' clubhouse. Bobby Thomson sits on a chair in front of his locker, champagne from a bottle an exuberant teammate has poured over his head dripping down his face. One of the myriad reporters surrounding him asks, "What kind of pitch did you hit?" "It looked like a high curve ball," Thomson replies, "I got around on it."

Neither my brief scene nor Hexter's effort to convey what he terms "a sense of vicarious participation in a great happening" captures the pain of losing, accentuated for me by memories of a boyhood largely spent rooting for the Dodgers, and surely as legitimate a perspective on history as the joy of winning. So we cross the way to the Dodger clubhouse, where, in heavy silence, Ralph Branca sits almost alone. The game is long over, the tears have dried on his face. Finally, one of the few reporters present risks a question: "What did you throw him?" "I hung a curve," Branca says.

Should the second scene be less revealing to the historian than the first? Hexter apparently thinks so. He seems to exclude the possibility that history, as Kenneth Burke pointed out, might be cast in a tragic, as readily as in a comic, mode. Arguing from what he labels Aristophanic principles, Burke associates tragedy with war and comedy with peace.[9] But he could as well (and more conventionally for a historian) have associated tragedy with losing and comedy with winning. Branca is said to have asked his parish priest, "Why me?" And despite the reassuring answer, "Because you were strong enough to bear the burden," he was never again the pitcher he had been in 1951. While the decline of a major league ballplayer's career hardly provides the stuff of Aristotelian or Shakespearean tragedy, it is consistent with the sort of tragic experience dramatized by, say, Eugene O'Neill, Tennessee Williams, or Arthur Miller.

Burke's "attitudes toward history" are partly defined by literary considerations—genre and tone—because for him, every event is "unique," resistant to the methods of science. Scientific history has, he observes ironically, yielded the realization that "gauging the 'right historical moment' [he is here primarily concerned with the right moment for enacting rather than writing history] is a matter of *taste*."[10] He would deny the existence of the "covering-law models" positivistic philosophers are fond of seeking, and sometimes find, for history, and would perhaps even demur at E. D. Hirsch's confidence that the meanings of texts (historical as well as literary) are "determinate and reproducible."[11]

Hexter clearly stands with Hirsch. He gleans the details surrounding Thomson's home run not from personal observation but from Russ Hodges's radio description—in effect a text, the meaning of which Hexter assumes he can accurately determine and reproduce. In stressing that he enjoys only indirect access even to an event contemporary with himself, and in which he had intense interest, Hexter is, moreover, reflecting the usual situation of the historian. For the historian is seldom witness to the history he writes; typically, he has reconstructed it from documents. He is, as Jay Cantor, extrapolating from the views of Claude Lévi-Strauss, suggests, something of a *bricoleur*, fashioning from the pieces he finds in history a kind of *bricolage*.[12] And he is almost always at the same remove from his subject as Hexter listening to Hodges, or as the reader of the book the historian's study of the documents has wrought.

Indeed, the transaction that took place between Hodges or his counterpart in the Brooklyn radio booth, Red Barber, and their audiences is analagous to the transaction that takes place between writer and reader. When their clubs were on the road, Hodges and Barber broadcast not from the stadium but from a studio, reading the action from Western Union tickertape. The tape carried only rudimentary data ("ball," "strike," "single to left"), which they elaborated imaginatively, drawing on their knowledge of the game to infer plausible contexts for the facts the tape reported. The listener, in turn, drawing on his knowledge, would transform their words into mental images.

This transaction parallels the transaction that, Barbara Herrnstein Smith suggests, takes place in reading, say, *Paradise Lost*. Milton, she argues, does not create Eve and Eden; he formulates statements about "Eve" and "Eden," thereby inducing the reader to create the character and setting for the story.[13] Thus no two Eves and Edens will be quite the same. Hans Robert Jauss develops a comparable view, specifically of literary history, which he claims traces through time the "realization of texts" by receptive readers as well as productive writers. To Jauss, literary history concerns not documentable facts, to be laid end to end in some sort of linear narrative, but an imaginative interaction between writers and readers that resists mere chronicling. Chrétien de Troyes's *Percival*, he argues, cannot be construed as historical in the way that the Third Crusade is historical; it cannot, that is, be accorded the status of a discrete event, caused by a complex of circumstances, both situational and psychological, and causing a series of further events.[14]

Might Jauss, however, be oversimplifying the problem of deciding what constitutes a "historical fact" or "event"? At this distance, do we comprehend the Third Crusade any differently than we do *Percival*? Can we know either as anything but the substance of narratives, texts to be interpreted? I do not suggest that we dismiss the Third Crusade, or any event for the reality of which we have sufficient evidence, as fiction. But we inevitably read narratives of events in ways analogous to the way we read fiction. As Fredric Jameson puts it, in a remark equally applicable to writers and to readers, we approach history, reality, through a refractive process he calls "prior textualization," "narrativization in the political unconscious."[15] Or as Jonathan Culler in a more ideologically muted statement formulates it, no reader opens a book a total innocent, a tabula rasa on whom the book's words are then inscribed. Reading is, no less than writing, "charged with artifice."[16] For readers bring to narratives, as writers to the act of narrating, assumptions, convictions, myths ("prior textualizations") by which they order, interpret, and evaluate language as well as life. Perceptive readers will disagree on the meanings of texts, as professional historians disagree on the meanings of events, or of the documents from which they reconstruct events. Meaning, as experience teaches us, cannot always be in Hirsch's sense determined.

How we respond to a "great happening" such as that in which Hexter invites us vicariously to participate depends on whose vantage we are induced, or disposed, to adopt: Hodges's?

Thomson's? Branca's? the ostensibly dispassionate historian's? Hexter would have us adopt the historian's, maintaining the appearance of dispassion by following Fischer's advice to exclude questions of why. The questions Hexter proposes for his historian are: "How did it come about that . . . ?" and "How did he (or they) happen to . . . ?"[17]

The questions my fictional reporters ask are similar: how? and, in the second scene, what? But to turn them, or those of Hexter's historian, into questions of why requires little manipulation, especially if one imagines the questioner to be not an interviewer querying a player but a scholar querying his evidence. In fact, Dodger manager Chuck Dressen was asked why: Why did he bring in Branca to pitch to Thomson?—a question I would judge to be perfectly proper, even crucial, for our baseball historian to pose. "Because [of the pitchers available] he had the best stuff," Dressen replied—an answer that, in its context, involves calculations as subtle, and to the baseball historian as significant, as those that brought Napoleon to Waterloo.

What we construe as history, then, depends on the interactions that take place, first between the historian and his evidence, second between the narrative he builds from the evidence and ourselves, his readers. The historical record, so called, is the outgrowth of a largely unwitting collaboration between those who, after the fact, sought to order and understand the materials of its making, to give them form, and later generations who have interpreted the texts their predecessors' understanding produced, and have fit that understanding into their own frames of reference. How we interpret the stories historians tell hinges on questions that not only they but we ask, and on the ways in which, the perspectives from which, both they and we ask them. As John Keegan observes about battles, the issues of who won, who lost, and why, which preoccupy commanders and the chroniclers of their exploits, differ from the concern of the line soldier, immersed in trying to comprehend what happened to him.[18] Leo Gershoy, biographer of Bertrand Barère, one of the real enigmas of the French Revolution, was I think right when, echoing R. G. Collingwood, he told a symposium on philosophy and history at New York University that the historian has virtually to merge his consciousness with the consciousness of his subject:

> Pragmatically speaking, historical facts, however numerous and detailed, did not take on significance for me until by using them and not ignoring them, I made them come to life. In that figurative sense I created them. Their existence could not be denied, but it was my selection that brought them into the world of history by lodging them in my mind for the purpose to which I intended to employ them.[19]

And Lee Benson was right when he told the same symposium that the phenomena historians study are too diverse, their aims in studying them too various, to be subsumed under general laws: "history is what historians do" (and, I would argue, what artists do when they re-create historical events).[20]

History is what Blake, Scott, Hazlitt, Carlyle, Hardy, and to a degree even Dickens supposed themselves to be "doing" in their portrayals of the French Revolution and the Napoleonic Wars. They were, moreover, not only writing kinds of history but, in the process, thinking about how history should be written, how it may be understood. They fixed on the French Revolution because they perceived it, for better or worse, to have been, in the words of Matthew Arnold, "a more spiritual event than our Revolution, an event of much more powerful and worldwide interest," and to have established "an order of ideas which are universal, certain, permanent."[21]

Whether they share Arnold's faith in Reason as manifesting this universality, certainty, and permanence (as most of them do not), Blake, Scott, Hazlitt, Carlyle, Dickens, and Hardy all see in the French Revolution a microcosm of total human experience, universal history. They all conclude that its progress cannot be simply reported, compiled in documented narratives stitching episodes end to end; it must be rendered, epitomized symbolically. Only by symbolic rendering can the depth, as well as the breadth, of its meaning be grasped. For them, the Revolution must be realized as myth, incorporating in its violent swings between the demonic and the divine the ultimate nature of man.

W. B. Yeats once remarked of Lord Acton that he claimed to

believe "in a personal devil, but as there is nothing about it in the Cambridge Universal History which he planned, he was a liar."[22] In calling up scenes of the Revolution or its aftermath, Blake, Scott, Hazlitt, Carlyle, Dickens, and Hardy would not be liars.

ONE

Fabricating History

By the River Chebar

"Reasons and opinions concerning acts, are," Blake proclaims in the *Descriptive Catalogue* to his exhibition of 1809, "not history. Acts themselves alone are history." Thus he announces himself a partisan of the movement against Enlightenment historiography in the then budding (and still flourishing) debate over how, or whether, the past can be plausibly represented to those living in the present. Acts, Blake adds, "are neither the exclusive property of Hume, Gibbon nor Voltaire, Echaid, Rapin, Plutarch nor Herodotus" (E, p. 534). Reasoning historians all, they twist cause and consequence; and in separating acts from their explanations—proposing chains of cause and effect to rationalize events, order narrative—they distort history.

Not by chance does Blake indict reasoning history in a catalog of pictures subtitled "Poetical and Historical Inventions." His pictures become both poetical and historical by incorporating the imagined into the actual, the general into the particular, the abstract into the concrete: by being rendered symbolic. Blake describes the works of the Asiatic Patriarchs he reports having seen in vision as "containing mythological and recondite meaning, where more is meant than meets the eye." Of Chaucer's Pilgrims, Blake observes that though their names and titles "are altered by time ... the characters themselves remain unaltered." They have "the physiognomies or lineaments of universal human life" (*DC*, E, pp. 523–24): they are simultaneously individuals and types.

Coleridge discerns these same physiognomies or lineaments in the figures he encounters in Scripture: "Both Facts and Persons," he insists in *The Statesman's Manual* (1816), "must . . . have a two-fold significance, a past and a future, a temporary and perpetual, a particular and a universal application. They must be at once Portraits and Ideals" (pp. 29–30). To be at once a portrait and an ideal is, in a phrase Blake could have supplied Coleridge, to embody "Divine Humanity" (p. 29).

The process by which humanity is affirmed as divine is epitomized, Coleridge suggests, in the wheels Ezekiel beheld while sitting among the captives by the River Chebar. History and political economy, whose eighteenth- and nineteenth-century permutations he dismisses (again Blake-like) as infected with "the general contagion of . . . mechanic philosophy," are in Scripture "living *educts* of the Imagination"; and by "living educts of the imagination" he means "that reconciling and mediatory power, which incorporating the Reason in the Images of the Sense, and organizing (as it were) the flux of the Senses by the permanence and self-circling energies of the Reason, gives birth to a system of symbols, harmonious in themselves, and consubstantial with the truths, of which they are the *conductors*" (p. 29). Incorporating reason in the images of sense, organizing the senses, entails projecting in microcosm the "radical monism" that, M. H. Abrams stresses, the Romantics (Coleridge prominently among them) drew from Plotinus and made the basis of their cosmology.[1] The images emanating from imagination that signify spiritual truth remain one with, that is, are consubstantial with, the truth they signify.

In this metaphysic, Abrams points out, lies the rationale for the circle described by the Greater Romantic Lyric. As Coleridge himself wrote in a letter to Joseph Cottle, "The common end of all *narrative*, nay of *all*, Poems is to convert a *series* into a *Whole*: to make those events, which in real or imagined History move in a strait *Line*, assume to our Understandings a *circular* motion—the snake with its Tail in it's Mouth." Or as Abrams (with a nod at Einstein?) interprets Coleridge's remarks, history is linear to the shortsighted but circular to the man "who Present, Past, & Future sees."[2]

Blake's Bard present, past, and future sees. And Blake too detects in the shape of history the snake with its tail in its mouth, perceived variously by "corporeal" eyes, accurately, univocally, only "in vision." Types of "all ages and nations," Chaucer's Pilgrims manifest the recurrence of the eternal in man. Though they may, age after age, assume appearances "different to mortal sight," they remain "to immortals only the same; for we see the same characters repeated again and again, in animals, vegetables, minerals, and in men." (*DC*, E, p. 523).

The problem that mortal eyes perceive askew what immortal eyes alone can see truly also vexes Carlyle. The historian, he concedes in "On History" (1830), manages merely "a poor approximation," at best glimpsing "mysterious vestiges of Him ... whom History indeed reveals, but only all History, and in Eternity, will clearly reveal" (*CME* 2: 89). Because he stands not at the end of history but within it, and thus knows nothing of the future, if something of the past, the historian is blind to what Louis O. Mink, borrowing from Boethius, labels the *totum simul*: the visionary's clear revelation of the God behind history, God's complete comprehension of the world in space-time, and the ideal configuration—universal history—to which, for Carlyle, as for Blake and Coleridge, narrative history had, however vainly, to aspire.[3]

The inescapable myopia Carlyle ascribed to historians continues to trouble both writers and readers of history. Entering, occupying, and leaving the world always, as Frank Kermode puts it, "in the middest," men are locked into narrow and necessarily distorted perspectives, which they seek to overcome by fabricating what Kermode calls "fictive concords"—causal sequences that proceed coherently from definite beginnings to definitive endings and enable men to claim order, significance for their lives.[4] Michael Ryan, in his attempt to wed Marx with Derrida, similarly claims something fictive (though hardly concordant) about narratives of events, which, he insists, can never be fully realized—in his Derridaean phrase, "made present" to the reader. The conclusion he draws from this insistence rebuts Kermode's. To Ryan, making sense of history requires eschewing beginnings and endings, translating increments of historical time into constructs other than dramatized "middests" that evolve from original crises and move toward ultimate resolutions.[5]

Though Carlyle's narrative practice aligns him much more nearly with Kermode's critical, and ideological, principles than with Ryan's, he too recognizes a fictive element, even in narratives that are seemingly factual. Recalling in "On History" Sir Walter Raleigh "looking from his prison-window, on some street tumult, which afterwards three witnesses reported in three different ways, himself differing from them all," he warns that "the most gifted man can observe, still more can record, only the *series* of his own impressions: his observation, therefore, to say nothing of its other imperfections, must be *successive*, while the things done were often *simultaneous*; the things done were not a series, but a group" (*CME* 2: 87–88). Carlyle's concern with the inability of narrative to represent events as simultaneously occurring is shared by Ryan, who complains that circumstances in history, made to seem one-dimensional on the printed page, invariably disclose multiple dimensions to the enlightened eye.[6] Ryan might, however, have been surprised to find his views anticipated not only by Carlyle but by Coleridge, developing the interdependence of metaphysics and poetic structure for Cottle, and dissecting rationalistic epistemology in *Biographia Literaria* (1817):

> Whenever we feel several objects at the same time, the *impressions* that are left (or in the language of Mr. Hume, the *ideas*) are linked together. Whenever therefore any one of the movements, which constitute a complex impression, are renewed through the senses, the others succeed mechanically. It follows of necessity therefore that Hobbs [*sic*], as well as Hartley and all others who derive association from the connection and interdependence of the supposed matter, the movements of which constitute our thoughts, must have reduced all its forms to the one law of time. (*BL*, pt. 1, p. 96)

The reduction of all forms of association to one law of time—series causally explained by documentary evidence—is precisely the constrictive mold into which Martin Heidegger, well over a century after Coleridge, accused rationalism, scientific method, of forcing historiography.[7] As early as 1793, Godwin had denied, if not the possibility, at least the probability of multiple causes as "contrary to the experienced operation of scientifical improvement," and described the universe as "a body of events in systematical arrangement," in which, for every human being, "there is a chain of events . . . going on in regular procession through the whole period of his existence."[8]

Carlyle, thirty-seven years later, set out to dismantle Godwin's model: "Narrative is *linear*, Action is *solid*. Alas for our 'chains,' or 'chainlets,' of 'causes and effects,' which we so assiduously track through certain handbreaths of years and square miles, when the whole is broad, deep Immensity, and each atom is 'chained' and complected of all!" (*CME* 2: 89). Rejecting chains of cause and effect as reductive, Carlyle, like Blake, demands of historians the what ("some picture of things acted"), implying that he can find the why and how himself. This insistence on what, to the exclusion of why and how, partly underlies Carlyle's famous definition of history as "the essence of innumerable Biographies" (p. 86). History is what men do.

Carlyle does not suggest that true history, universal history, need consist of as many histories as there are men doing—that the field of history must be construed metonymically, to borrow an epithet from Hayden White. History is also what God writes. Carlyle, that is, offers no aid or comfort to Ryan, who reduces all discourse to images of images, departures from the reality they would evoke, and thus all efforts at historical reconstruction to stages in an infinite regress. Nor does he support Michel Foucault, who replaces universal history ("total history" as he terms it) with "general history"—in effect, histories.[9] To Carlyle, history is finally a "complex [and coherent] Manuscript," a "*Palimpsest*" of prophecies, "still dimly legible" and in fragments decipherable (p. 89).

His analogy of history to a palimpsest anticipates Derrida's of language, in dream and discourse, to inscriptions on a child's toy, the "Mystic Writing-Pad," which leaves traces of what has been erased detectable beneath the fresh figures drawn on its surface.[10] But to Derrida, these traces are presences only in the minds of their perceivers, shadows of shadows, reflections of reflections; to Carlyle, they are presences in the world, coalescing into tangible unity in the mind of God.[11]

White identifies the mode of Carlylean history as metaphor, synthesis, though he adds that Carlyle reacts against Enlightenment skepticism by discovering meaning to human life wholly in human life.[12] Yet if the record of human life is inscribed on a palimpsest of prophecies, its meaning must be traced ultimately to Providence. Carlyle shares the general Romantic perception of history as a drama by a hidden dramatist, who promises a millennial denouement.[13] This drama is the fictive concord Kermode attributes to the yearning for total order human beings inherit as denizens of an uncertain "middest." What makes Kermode's fiction concordant is the inevitable complementarity it projects between origin and end, creation and apocalypse.[14]

The middest, everything between origin and end, is determined by that complementarity. Indeed, origin and end are embedded in the events that comprise the middest, which is why Carlyle can claim that, in the palimpsest, "some letters, some words, may be deciphered; and if no complete Philosophy, here and there an intelligible precept, available in practice, be gathered" (CME 2: 89–90). Or as he declares in his afterthought to "On History," entitled "On History Again" (1833), written as the idea for an epic about the French Revolution was beginning to grip him, "History is the Letter of Instruction, which the old generations write and posthumously transmit to the new" (CME 3: 167).

History, that is, as Carlyle iterates and reiterates in his meditations on the uses of the past, "is philosophy teaching by experience." This conception of history links him to the Enlightenment historiography he professed to loathe. He has quietly, and with only slight alteration, lifted his aphorism from Lord Bolingbroke, who in his *Letters on the Study and Use of History* (1752) defines history as "philosophy teaching by example." Carlyle shares the conviction that history teaches not only with Bolingbroke but with Edmund Burke. "In history," Burke declares, "a great volume is unrolled for our instruction, drawing the materials of future wisdom from the past errors and infirmities of mankind" (RRF, p. 173).

Burke's didacticism reflects a view of history that Carlyle's exact contemporary, Leopold von Ranke, was firmly, if gently, to repudiate in the preface to his study of the early Latin and

Germanic nations: "To history has been assigned the office of judging the past, of instructing the present for the benefit of future ages. To such high offices this work does not aspire: It only wants to show what actually happened."[15] But judging the past and instructing the present sum up the office to which William Gordon aspires in his *History of the Rise, Progress and Establishment of the Independence of the United States of America* (1788), a work that boasted among its subscribers Richard Price and one William Blake, Esq. "It should oblige all, who have performed any distinguished part on the theatre of the world," Gordon announces in his preface, "to appear before us in their proper character; and to render an account of their actions at the tribunal of posterity, as models which ought to be followed, or as examples to be censured and avoided."[16]

Theater of the World

Gordon's trope of history as a play, ongoing from eternity to eternity in the theater of the world, seems also to be the trope implicit in Blake's parade of Chaucerian characters, in each age appearing different to mortal eyes but to immortal eyes always the same. Those immortal eyes looking down on the stage of history will a century later have become the Spirit audience at Hardy's vast "historical Drama," *The Dynasts* (1904–1908).[17] As Frances Yates points out, analogizing the world to a cosmic theater was a rhetorical stratagem widely employed in the Renaissance: "All the world's a stage / And all the men and women merely players."[18] This analogy reappears in the biographies of Napoleon by Scott and Hazlitt. The battle of Leipzig is, for Scott, "the last act of the grand drama, so far as the scene lay in Germany" (3: 57). The return from Elbe is an "extraordinary drama," in which Napoleon plays "the part destined for him" (3: 201), presumably by the hidden dramatist. The allies' entrance into Paris in 1814 is, for Hazlitt, also a "drama," which might have proved "tragical" but turns out merely "sentimental."

Hazlitt ascribes the sentimental (read "farcical") turn taken by this drama to France's refusal to play its proper role: "The French are a people who set almost as much store by words as by things, and who very much prefer the agreeable to the disagreeable: they therefore took the word of the allies that nothing was meant but to oblige" (15: 200). Indeed, he and Scott, who almost never agree about Napoleon, agree that the instinct of his subjects is to eschew any role at all, to behave not as actors but as audience. Scott explains the fervor to which Napoleon's rhetoric raises his troops as a manifestation of French national character: "they are willing to take everything of a complimentary kind in the manner to which it seems to be meant. They appear to have made that bargain with themselves on many points, which the audience usually do in a theatre,—to accept the appearance of things for the reality" (1: 312). Hazlitt explains the Parisians' docility before the allied invasion, and despite Napoleon's heroic resistance, by their detachment: "They had been accustomed to sit, as in a theatre, and enjoy the roar of victory at a safe distance; but when this grand drama of war was turned to serious earnest and brought home to themselves, they did not at all know what to make of it" (15: 174).

Both Scott and Hazlitt seem to detect in French conduct a glimmer of the psychology, explored by Paul Fussell, that disposed soldiers in the Great War to treat warfare as theater—to perceive themselves as half performers, half spectators, thereby clinging with at least part of their minds to the conviction that their "real" selves could remain separate from the madness around them.[19] Citing the delusion of a patient who thought himself invisible, R. D. Laing remarks that in a dangerous environment, invisibility makes the best defense.[20] In no environment do men situate themselves less visibly than on a battlefield—except perhaps in the seats of a darkened theater.

Burke and Coleridge, if framing their metaphors differently, also react to a perceived threat in their environment (which they hear resonating to "Ca Ira" rather than to "God Save the King") by rendering events as dramatized scenes. Coleridge, who in 1795 still thought Robespierre a better man than Pitt, depicts the Incorruptible in *Conciones ad Populum or Addresses to the People* as having—his conscience muted by the adulation of the mob—"masqueraded on the bloody stage of Revolution, a Caligula with the cap of Liberty on his head" (1: 35). Five years earlier, Burke, focusing not on an individual but on the mass, had labeled 1789 a "monstrous tragi-comic scene," in which "the most opposite passions necessarily succeed, and sometimes mix with each other in the mind; alternate contempt and indignation; alternate laughter and tears; alternate scorn and horror" (*RRF*, p. 9).

Although Burke adds that in some Frenchmen, his real villains, "this strange scene inspired . . . [only] exultation and rapture" (*RRF*, p. 9), his account of Parisians provoking, and provoked by, wrenches between opposite passions suggests the schizoid conduct Masao Miyoshi discerns in the villains of Gothic romance, whom he pronounces an "archetype for alienated man divided against himself."[21] Those Gothic villains were intended, repeatedly in the late eighteenth and early nineteenth centuries, to induce the awe and terror Burke had pronounced sublime. In his *Philosophical Enquiry into the Origins of our Ideas of the Sublime and Beautiful* (1757) he concedes, however, that the ultimate vehicle for inducing awe and terror—the ultimate drama—is reality itself:

> Chuse a day on which to represent the most sublime and affecting tragedy we have; appoint the most favorite actors; spare no cost upon the scenes and decoration; unite the greatest efforts of poetry, painting and music; and when you have collected your audience, just at the moment when their minds are erect with expectation, let it be reported that a state criminal of high rank is on the point of being executed in the adjoining square; in a moment the emptiness of the theatre would demonstrate the comparative weakness of the imitative arts, and proclaim the triumph of the real sympathy.[22]

Burke's attacks on the French Revolution reflect his struggle to prevent the triumph of the real sympathy in England. Englishmen debating the revolution in France are weighing revolution themselves.[23] Watching the scenes unfold across the Channel, they are like a theater audience, controlled by what Herbert Lindenberger calls the audience's "double view": they sympathize with the actors in the conflict, even while keeping their distance, as from "mere actors" in a play.[24] Burke and Coleridge deploy their rhetoric as a force to hold their compatriots where they are, to restrain them from abandoning the play for the streets. They argue, at least metaphorically, that the violence sweeping France is itself somehow less than real, and that in any case, England had already confronted that violence, and domesticated it, in the revolution of 1688.

The insidious attraction of this argument led Thomas Paine to assail Burke, in the *Rights of Man* (1791), for reducing history to theater. Of Burke's solicitude towards the Bourbons, Paine observes that "his hero or his heroine must be a tragedy-victim, expiring in show, and not a real prisoner of misery, sliding into death in the silence of a dungeon" (pp. 288–89). Of Burke's emphasis on the havoc wreaked by those prisoners of misery gathered into a mob ("a swinish multitude" [*RRF*, p. 95]), Paine observes that "It suits his purpose to exhibit the consequences without their causes. It is one of the arts of the drama to do so" (*RM*, p. 297).

Exhibiting consequences without causes is precisely the indictment one of Blake's reasoning historians, or Godwin, might have brought against Burke. Exhibiting a cause for every consequence, however, is precisely the indictment Coleridge brings against reasoning history and its kindred discipline, politics:

> The rational instinct, therefore, taken abstractedly and unbalanced, did *in itself*, ("ye shall be as gods!" Gen. iii. 5) and in its consequences, (the lusts of the flesh, the eye, and the understanding, as in verse the sixth,) form the original temptation, through which man fell: and in all ages has continued to originate the same, even from Adam, in whom we all fell, to the atheist who deified the human reason in the person of a harlot during the earlier period of the French Revolution. (*SM*, pp. 61–62)

Typical of Christian exegetes in the eighteenth and nineteenth centuries, as Abrams points out, Coleridge discerns in the events that constitute the "middest" refashionings or foreshadowings of the beginning and the end.[25] The Revolution and Napoleon at last behind him, he prescribes, as an "antidote" to what he can in retrospect dismiss as "mere restless cravings," a thoughtful assimilation of "the events of our own age to those of the time before us" (*SM*, p. 9). By "the time before us," he means not only the arduous struggle against Napoleon and the "*monstrum hybridism*" of Jacobinism (*SM*, p. 65), but England's banishment of James II, conducted under the "especial controul of Providence to perfect and secure the majestic Temple of the British Constitution!" (p. 109), and those other episodes conducted under the special control of Providence and contained in Scripture:

> If this [distillation of the antidote to human unrest] be a moral advantage derivable from history in general, rendering its study therefore a moral duty for such as possess the opportunities of books, leisure and education, it would be inconsistent even with the *name* of believers not to recur with pre-eminent interest to events and revolutions, the records of which are as much distinguished from all other history by their especial claim to divine authority, as the facts themselves were from all other facts by especial manifestation of divine interference. (*SM*, p. 9)

The records of those events and revolutions on which Coleridge would have believers focus their preeminent interest are his equivalent to the palimpsest that carries Carlyle's letters of instruction to future generations. Napoleon betrays "the Marks, that have characterized the Masters of Mischief, the Liberticides, and mighty Hunters of Mankind" (*SM*, p. 65) back to Nimrod and eventually to Satan. England's failure to move decisively against Napoleon before he had solidified his power is, to Coleridge looking back, an outgrowth of a failure to heed the instructions the letters on the palimpsest conveyed: "I well remember, that when the examples of former Jacobins, as Julius Caesar, Cromwell, and the like, were adduced in France and England at the commencement of the French Consulate, it was ridiculed as pedantry and pedant's ignorance to fear a repetition of usurpation and military despotism at the close of the ENLIGHTENED EIGHTEENTH CENTURY" (*SM*, p. 11).

What Coleridge is proposing, despite his discomfort with the idea, is that history, to be understood as prophecy, must be interpreted as allegory. Reading Ezekiel's river as "the stream of time," he insists that from "our relative position on its banks" (*SM*, p. 29), our ability to see past and future embedded in the present depends on our ability to recognize in the actions of the present the transgressions of the past. "A symbol," he argues, in an effort to clarify the distinction between symbolism and allegory, incorporates "the translucence of the Eternal through and in the Temporal. It always partakes of the Reality which it renders intelligible; and while it enunciates the whole, abides itself as a living part of the Unity, of which it is representative" (*SM*, p. 30). Napoleon is both himself, partaking of, indeed largely shaping the reality of the fifteen-year struggle between imperial France and the rest of Europe, and the embodiment of a tyrannical will whose archetype is the pride of Satan warring on God.

History, then, is analogous to, if not identical with, the Vitruvian memory theater Yates reconstructs in *Theatre of the World*. Man is a microcosm who plays out his role in a setting designed (as the "Globe" in Globe Theatre implies) to evoke the macrocosm.[26] Jules Michelet suggests as much. He perceives in the French Revolution the pattern of a morality play—"I see upon the stage but two grand facts, two actors and two persons, Christianity and the Revolution"—and nurtures the germ to become his *History of the French Revolution* (1847–1853) by meditating on his own life: "I commence with my own mind. I interrogate myself as to my teaching, my history, and its all-powerful interpreter,—the spirit of Revolution. . . . In it alone France becomes conscious of herself."[27]

Different Stories

Michelet's testimony to the interaction of self and circumstance as basic to the process of writing his *History* anticipates R. G. Collingwood's proposal that the interaction of self and circumstance is basic to the writing of all history. The historian, Collingwood declares, comprehends history by reliving in his mind the feelings, thoughts, and deeds of the actors in the world he re-creates.[28] Understanding, as Wilhelm Dilthey represents it, consists of rediscovering "the I in the Thou" and, in what amounts to an almost mystical expansion of consciousness, extending the rediscovery to "ever higher levels of connectedness" until consciousness embraces "the totality of mind and universal history."[29]

Hazlitt, thinking more about the future than the past, had by 1805 already grasped this notion of how the historian understands. Probing the psychic forces that energize men's growth into moral beings in *An Essay on the Principles of Human Ac-*

tion, he announces that "the imagination ... must carry me out of myself into the feelings of others by one and the same process by which I am thrown forward as it were into my future being, and interested in it" (1: 1–2). David Bromwich claims that Hazlitt's case for empathy as the mind's mode of understanding (turned by Keats, who attended some of his lectures, into the concept of "negative capability") influenced not only the aesthetics but the politics of both Hazlitt and Keats. If the purpose of literature is to induce the reader to transgress the bounds of self and to enter alternative, perhaps alien, selves, the bounds imposed by society should be no less transgressible.[30]

Even Scott, who in his *Life of Napoleon Buonaparte* (1827) affirms as sacrosanct bounds imposed by society, perceives in his narrative a vehicle for transgressing bounds of time and (implicitly) space:

> That which is present possesses such power over our senses and our imagination, that it requires no common effort to recall those sensations which expired with preceding events. Yet to do this is the peculiar province of history, which will be written and read in vain, unless it can connect with its details an accurate idea of the impression which these produced on men's minds while they were yet in their transit. (1: 17)

Conveying accurately the impressions events have produced on men's minds is, to Scott, largely a task for the historical novel. As Avrom Fleishman observes in *The English Historical Novel*, in which Scott figures as one of the genre's main practitioners, fiction illuminates the past by enabling us to see ourselves in it.[31] Or in Robert Langbaum's description of Romantic epistemology in general, the artist apprehends the experience of the "not-me" by playing the roles the "not-me" has lived.[32]

In playing the "not-me, " and especially in calling up beings from the past, the artist manifests a dual consciousness. He is both his character enmeshed in circumstances and the historian who looks back on circumstances and sees them whole. In the French Revolution specifically he is also, as Jean-Paul Sartre argues, both an actor of real circumstance, engaged in revolutionary struggle, and an actor playing an actor—insofar as the Revolution found its model in republican Rome, a Girondin seeking to revive Cato in himself. "Imagine," Sartre asks his reader, "an actor . . . playing Hamlet"; whereupon he proceeds to explore the complexities of awareness the actor brings with him to the theater:

> He crosses his mother's room to kill Polonius hidden behind the arras. *But that is not what he is actually doing.* He crosses a stage before an audience and passes from "court side" to "garden side" in order to earn his living, to win fame, and this real activity defines his position in society. But one cannot deny that the movement of the imaginary prince expresses in a certain indirect and refracted manner the actor's real movement, nor that in the very way in which he *takes himself* for Hamlet is his own way of *knowing himself* an actor.[33]

Sartre claims to be extending the Marxist metaphor of man's relation to history "to its limit": we grasp the past by reliving it, and thus become alert to our power in the present; we learn not merely how to interpret the world but how to change it.

The changes Sartre perceives the Girondins effecting were, the Marxist side of himself concludes, truncated by the myopia of their class: "Their way of *calling* themselves Cato is their way of *making* themselves bourgeois, members of a class that discovers History and that already wants to stop it."[34] To stop history requires the capacity to evoke from the past an ideal, which is then established in the present—the capacity to realize what a Marxist historian of the Revolution calls the struggle's "mythic character."[35]

Sartre's absorption in this "bourgeois" effort, through a kind of imaginative leap, to realize the Revolution's mythic character troubles Fredric Jameson, who finds Sartre's apparent aim, to relive the Girondins' experience, akin to the aim of a historical novelist.[36] Sartre, he seems to suggest, understands the historian's task as Hexter understands it: to draw the reader—even as a historical novelist might—into vicarious participation in a great happening. To the Marxist ideologue, vicarious participation is insufficient, perhaps counterrevolutionary: it encourages quiescence, detachment. Although narratives, historical and fictional alike, proceed episode by episode, com-

131/925

municating to us some of the uncertainty of life, they are almost invariably told in the past tense.[37] They treat events as finished, thus distancing us from their reality. They are Aristotelian constructs, with beginnings, middles, and especially ends.

White stresses the degree to which historical narratives are constructed, labeling them "verbal fictions," closer to literature than to science.[38] He means what Paul Ricoeur means when he asserts that time becomes human time by being plotted, given significant form.[39] Both White and Ricoeur are, with appropriate caution, expressing the kind of skepticism toward history expressed by Claude Lévi-Strauss, who first defines historical fact conventionally as what happened, then asks where it happened. The past exists for us, he proposes, necessarily refracted by preconceptions that are cerebrally, even chemically, determined. Written history, he insists, is "never history, but history-for."[40] And historical fact is never, as scientific fact may often be, divorced from complicating contexts, whether geographical, political, or metaphysical.[41] As Kenneth Burke remarks, Aquinas acknowledged the division of peoples into classes (with its attendant consequences—government, property, slavery) as an outgrowth of the Fall, to be expunged only at the millennium; Marx acknowledged the same phenomenon as an outgrowth of capitalism, to be expunged by a popular revolution.[42]

To write "history-for," then, is to shape it in accordance with a teleological, if not an ideological, design, which is partly why White likens historical narratives more to literature than to science. In this sense, however, science too may be more literary than scientific. John C. Greene finds it no coincidence that the advocates of natural selection during the first half of the nineteenth century were British to a man, all products of a society in which competition and survival of the fittest were assumed to fuel progress.[43] And Collingwood applies this unscientific, or at least unpositivistic, notion of science to the scientist's way with language. Citing the reflex that causes the hedgehog, when threatened, to roll itself into a ball, he observes that scientists describe such behavior as a "defensive mechanism." But "defensive mechanism," he insists, can only be a metaphor, "for a mechanism means a device, and a device implies an inventor."

He views the language of historians comparably, declaring that "we cannot think about history without using similarly teleological metaphors." To show how we commonly think about history, he summarizes a standard approach to representing the spread of Roman power:

> We use phrases like the conquest of the Mediterranean world by Rome; but actually what we mean by Rome is only this and that individual Roman, and what we mean by the conquest of the Mediterranean world is only the sum of this and that individual piece of warfare or administration which these men carried out. None of them actually said "I am playing my part in a great movement, the conquest of the Mediterranean world by Rome," but they acted as if they did say that, and we, in looking at the history of their actions, find that these actions can only be envisaged as if they were controlled by a purpose to achieve that conquest, which, as it certainly was not the purpose of this or that individual Roman, we metaphorically describe as a purpose of nature.[44]

Collingwood takes an agnostic—which is to say a particular ideological and hardly inevitable—stance toward history and science. Whether we could imagine an ancient Roman claiming to have played his part in a great movement, Rome's conquest of the Mediterranean world, we could, I suspect, easily imagine that a nineteenth-century Englishman not only might make this claim about his part in England's conquest of some corner of the world but might also justify it by appeal to a purpose higher than Whitehall's. William R. Greg, in an article written for *Fraser's Magazine* in 1868, and read by Darwin, sought to ease whatever doubts his readers may have had about the morality of exploiting, even exterminating, weaker peoples for profit by pronouncing such practices consistent with natural selection: "The process is quite as certain, and nearly as rapid, whether we are just or unjust; whether we use carefulness or cruelty. Everywhere the savage tribes of mankind die out at contact with the civilized ones."[45] In extending its power over what were perceived as the desolate places of the earth,

England merely acts according to natural law; and natural law easily translates into divine right. Convinced of the moral superiority of empire, Paul Wilcox, in *Howards End* (1910), goes to Nigeria to carry the burden of civilization.

Civilization also constitutes the burden Scott sees England carrying in the Napoleonic Wars. He discerns in the French Revolution a reenactment of the Fall, and in England's struggle against Napoleon God's means of setting Europe back on its proper course. Dickens reduces England to one Englishman, Sydney Carton, making his thoughts, as he awaits the guillotine, a prophecy of redemption for France and of an idyllic future in England for the Darnays, both the results of his Christlike sacrifice. Even Hardy, who shares Collingwood's unease with history teleologically rendered, announces as his intention in *The Dynasts*, to glean from the Napoleonic Wars a heroic myth for Albion: "The slight regard paid to English influence and action throughout the struggle by so many Continental writers who had dealt with Napoleon's career, seemed always to leave room for a new handling of the theme which should re-embody the features of this influence in their true proportion" (pp. xxiii–xxiv).

Weighing English influence in its true proportion became a matter of dispute not only on the Continent but in England as well. Although Blake and his fellow radicals, Richard Price and Joseph Priestley, share Scott's providential view of history, they detect in the French Revolution itself, rather than (as Scott does) in its defeat, a foreshadowing of the millennium. Although another of Blake's radical compatriots, Paine, shares, with Hazlitt, Hardy's agnostic view of history, both Paine and Hazlitt rage at policies eventuating, for Hardy, in an epic drama of (mainly) English arms and men.

That their rage is repeatedly aimed at Edmund Burke reflects his prominence as a spokesman for Toryism. It also attests to the astuteness of Lévi-Strauss's remark that history written is "history-for." Despite Burke's secure place in the pantheon of British culture, he has come under attack too from a modern American historian with strong liberal sympathies, R. R. Palmer. Arguing that Burke embraced Tory values in reaction neither to the doctrinaire rationalism of the Enlightenment nor to the excesses of 1789, but that he had announced his allegiance to these values much earlier in debates over social policy in Britain for Britons, Palmer labels Burke's stature as a critic of revolution "entirely a later concoction," and adds that his outbursts "dismayed" his contemporaries.[46]

By contemporaries, Palmer means Burke's colleagues in Parliament, allies as well as opponents. One of his (admittedly younger) contemporaries outside Parliament, however, had recognized the same political consistency in Burke to which Palmer calls attention, but with very different feelings. Coleridge in *Biographia Literaria* invites the scholar who doubts the virtue of "unanimity," and its source in principle, to examine "the speeches and writings of EDMUND BURKE at the commencement of the American war, and compare them with his speeches and writings at the commencement of the French war. He will," Coleridge assures the scholar, "find the *principles* exactly the same and the deductions the same; but the practical inference almost opposite, in the one case, from those drawn in the other; yet in both equally legitimate and in both equally confirmed by the results."

In this consistency of principle Coleridge discovers the key to Burke's greatness: "How are we to explain the notorious fact that the speeches and writings of EDMUND BURKE are more interesting at the present day, than they were found at the time of their first publication; while those of his illustrious confederates are either forgotten or exist only to furnish proofs, that the same conclusion, which one man had deduced scientifically, *may* be brought out by another in consequence of errors that luckily chanced to neutralize each other." By crediting Burke with having deduced conclusions scientifically, Coleridge means that Burke "referred habitually to *principles*"; and his habitual reference to principles made him a prophet: "For every *principle* contains in itself the germs of a prophecy; and as the prophetic power is the essential privilege of science, so the fulfillment of its oracles supplies the outward and (to men in general) the *only* test of its claim to the title" (*BL*, pt. I, pp. 191–92).

The Bourbon monarchy restored, in his view permanently, Coleridge lauds Burke's prophetic power, unaware (as Palmer, or Marx, might well have stressed) of the upheavals that 1830 and 1848 held in store. His understanding of the French Revolution, like theirs, is conditioned by the values he brings to it and the point in time from which he apprehends it. Historians, as Dominick La Capra notes, do not always know how the story turns out.[47]

Historical knowledge, then, seems as Ricoeur defines it: a limited vision, rendered meaningful by conformity to some causal or teleological model, and incorporating the preconceptions of the modelmaker.[48] Thus narratives that convey historical knowledge may be, as Collingwood comes close to suggesting, and as Roland Barthes specifically asserts, no more than linguistic constructs, in White's phrase, "verbal fictions."[49] Coleridge, who, like Carlyle, had already wrestled with most of the problems that occupy modern critical theorists, acknowledges the dilemma attached to claiming literal truth for historical narrative:

> The hypothesis of an external world exactly correspondent to those images or modifications of our own being, which alone . . . we actually behold, is as thorough idealism as Berkeley's, inasmuch as it equally (perhaps, in a more perfect degree) removes all reality and immediateness of perception, and places us in a dream-world of phantoms and spectres, the inexplicable swarm and equivocal generation of motions in our own brains. (*BL*, pt. 1, p. 137)

The qualities of dream are precisely those that White ascribes to the ostensibly factual surface of historical narrative. The methods by which Ranke and his followers purport to build their narratives from images exactly correspondent to the external world, to show what actually happened, or by which the positivists purport to extrapolate from what actually happened laws that govern the dynamics of history, amount to the process White calls (with a nice play on Freudian terms) "knowledgework": the historian "condenses" and "displaces," choosing from the materials before him details to include or exclude, emphasize or deemphasize, attach or detach; he encodes some of these details as causes, others as effects; and he reinforces his claim to scientific accuracy by weaving into his work a "secondary elaboration," a commentary "rationalizing" the structure of his story.[50]

At best, this process yields, in Barthes's estimation, the illusion of referentiality. As a consciousness speaking from a particular vantage, the narrator appears to be effaced; his words appear, like the frames of a film, simply to image the events they evoke. But in fact, Barthes would argue, the narrator's words refer to nothing beyond themselves.[51] White is suggesting, however, not that the historian's words are divorced from all referents but that, like the images of a dream, they distort their referents, though in a manner that the historian labors to resist. As "dreamwork" seeks to hide the latent meaning of dreams by obscuring their manifest content, "knowledgework" seeks to reveal the latent meaning of history by emphasizing its manifest content.[52] White shares Coleridge's insistence that, despite our need to rely on apparatus that refracts the images of the world we perceive, "Knowledge without correspondent reality is no knowledge." Knowledge grows, Coleridge further argues, from our search for correspondent reality: "To know is in its very essence a verb active" (BL, pt. 1, p. 164).

Ways of Knowing

To read is also, Ricoeur reminds us, a "verb active." When we read a story, we actualize it; in actualizing it, we reconfigure the experience it has related in words.[53] Reading becomes a way of knowing, and for Ricoeur too, what we know corresponds, however imperfectly, to what is. He is fighting the same battle fought by E. D. Hirsch and Gerald Graff, against those radical skeptics (such as Barthes or Derrida) who deny the possibility of objective meaning in language and whom Hirsch lumps under the memorable phrase "cognitive atheists." While Graff considers naive the nineteenth-century faith in the observer's ability to efface himself, to reduce himself to an instrument for recording raw facts, he pronounces "observational neutrality"—objectivity—"approachable as an ideal." Hirsch agrees: "The reader should try to reconstruct authorial meaning, and he can in principle succeed."[54]

No cognitive atheist would have missed the signs of retreat implicit in formulations like "as an ideal" and "in principle." To aim at success "in principle" is to acknowledge failure in practice. Graff and Hirsch appear to have left their campfires burning to fool the patrols and to have abandoned the field. In proposing, as the intent of narrative, reconfiguration rather than reconstruction, however, Ricoeur has established a new, and defensible, line. To reconfigure a story is to render it, if not factually accurate, at least analogous to life—to allow the possibility of a distinction between facts and truth. Analogousness to life is a feature J.R.R. Tolkien ascribes even to the most fabulous narratives: in his essay, "On Fairy Stories," he asserts the kinship of history and myth, for "both [are] ultimately of the same stuff"; in a foreword to *The Lord of the Rings*, he characterizes his own gigantic attempt at mythmaking as "feigned history."[55]

Scott had recognized romance as feigned history, adjudged history and myth to be ultimately of the same stuff, as early as 1824, in his "Essay on Romance," where he traces romance and "real history" to a "common origin" and constructs a hypothesis to account for their divide:

> The father of an isolated family, destined one day to rise into a tribe, and in farther progress of time to expand into a nation, may, indeed, narrate to his descendants the circumstances which detached him from the society of his brethren, and drove him to form a solitary settlement in the wilderness, with no other deviation from truth . . . than arises from the infidelity of memory, or the exaggerations of vanity. But when the tale of the patriarch is related by his children, and again by his descendants of the third and fourth generation, the facts it contains are apt to assume a very different aspect. The vanity of the tribe augments the simple annals from one cause—the love of the marvellous, so natural to the human mind contributes its means of sophistication from another—while, sometimes, from a third cause, the king and the priest find their interest in casting a holy and sacred gloom and mystery over the early period in which their power arose. And thus altered and sophisticated from so many different motives, the real adventures of the founder of the tribe bear as little proportion to the legend recited among his children as the famous hut of Loretto bears to the highly ornamented church with which superstition has surrounded and enchased it.[56]

Scott's theory of the origin of romance—derived from his speculations on how the story of Abraham and his descendants evolved into the history of the Jews—is exactly reversed by Northrop Frye, who theorizes that romance formed from a process not of mythicizing human experience but of humanizing mythic experience.[57]

Frye's theory, directed toward establishing the place of romance in the cycle of genres that constitute literature as a whole, essentially repeats Arnold Toynbee's theory of the origin of history and its place in the range of narrative types. To Toynbee, history may be symbolically shadowed in myth, "a primitive form of apprehension and expression in which . . . the line between fact and fiction is left undrawn."[58] Building on this thesis, he considers as alternative sources for the impetus behind history individual genius or impersonal movements:

> It is clear that if the geneses of civilizations are not the result of biographical factors or of geographical environment acting separately, they must be the result of some kind of interaction between them. In other words, the factor which we are seeking to identify is something not simple but multiple, not an entity but a relation. We have the choice of conceiving this relation either as an interaction between two inhuman forces or an encounter between two superhuman personalities.[59]

E. H. Carr poses a comparable choice. If he does not quite embrace an alternative definable simply as inhuman forces, he unhesitatingly dismisses the alternative of superhuman (or even human) personalities, relegating it to "primitive stages of historical consciousness": Wat Tyler and Pugachev have survived in history as rebels against oppression because of the masses united behind them; they "are significant as social phenomena, or not at all."[60]

Carr takes a Marxist view of how history happens. In *The Eighteenth Brumaire of Louis Bonaparte* (1852, 1869), Marx himself protests the magnification of a petty despot, inherent (he claims) in the treatments of the coup by Hugo and Proudhon, emphasizing instead *"class struggle"* as the force that enabled "a grotesque mediocrity to play a hero's part" (p. 8). This tendency to submerge the individual in the mass is, as Jameson points out, one of the grounds of Sartre's quarrel with orthodox Marxism. Seeking reasons for the National Assembly's declaration of war against the First Coalition in 1792, Sartre rejects the Marxist argument that the Assembly was manipulated by a commercial bourgeoisie scheming to increase profits, because it requires "those men whom we know well [Brissot, Gensonné, Vergniaud] to disappear . . . or else it constitutes them, in the final analysis, as the purely passive instruments of their class."[61]

Sartre displays an essentially literary sensibility toward history. If individuals—characters—disappear, or become mere passive instruments, narrative (at least in the traditional sense) hardly seems possible. Indeed, Carr excludes any notion that history may be a partly symbolic, and thus literary, form of narrative. Literature he defines as "a telling of stories and legends without purpose or significance."[62] He draws a clear line between fact and fiction, and assumes, as Toynbee does not, that history can keep to the factual side of the line.

Toynbee does not even suppose that history should keep to the factual side of the line. Of the alternatives he explores as routes to comprehending the forces that engender civilizations, he invites his reader to join him on ("yield our minds to") the second, traced in myths of encounters between superhuman personalities. "Perhaps it will lead us towards the light."[63] Whereupon he surveys a spectrum of such encounters—Yahweh and the Serpent in Genesis, the same antagonists transfigured in the Gospels, God and Mephistopheles in *Faust,* Artemis and Aphrodite in *Hippolytus*—whose conflicts have produced some of the greatest dramatic moments in literature and (as Toynbee interprets them) symbolize crucial turning points in history.

Literature includes, for him, a telling of stories and legends of the highest purpose and significance. He reads literature analogically, symbolically—as Blake reads *The Canterbury Tales:*

> Visions of these eternal principles or characters of human life [portrayed by Chaucer] appear to poets, in all ages; the Grecian gods were the ancient Cherubim of Phonecia; but the Greeks, and since them the Moderns have neglected to

subdue the gods of Priam. These Gods are visions of the eternal attributes, or divine names, which, when erected into gods, become destructive to humanity. They ought to be the servants, and not the masters of man, or of society. They ought to be made to sacrifice to Man, and not man compelled to sacrifice to them; for when separated from man or humanity, who is Jesus the Savior, the vine of eternity, they are thieves and rebels, they are destroyers. (DC, E, p. 527)

Embedded in the *Tales* is an allegory of universal history: "Every age is a Canterbury Pilgrimage" (DC, E, p. 526). And implicit in every Canterbury Pilgrimage is a version of the history, epitomized in *The Marriage of Heaven and Hell* (1793), of man's lapse from the Human Form Divine to the state of priest-ridden slave, who has forgotten "that all deities reside in the human breast" (pl. 11; E, p. 37).

Frye, whose theory of genres seems to have emerged from his study of Blake, proposes a model for romance that, in its darkest parodic form, fashions it into a similar kind of allegory. Frye's romantic world is demonic, dominated by a Urizenic "tyrant-leader, inscrutable, ruthless, melancholy, and with an insatiable will," his ego colossal enough to encompass "the collective ego of his followers." That ego, and those of his followers, must, in the scheme Frye develops, be fed by the murder of the tyrant's opposite (or perhaps rival), who thereby comes to represent a *pharmakos*, the sacrificial victim of myth.[64]

Although Frye disclaims any attempt to interpret literary narrative as a symbolic version of history ("The poet," he asserts, "never imitates 'life' in the sense that life becomes anything more than the content of his work"[65]), the pattern he finds in romance compellingly suggests the French Revolution. Ronald Paulson observes that the Revolution can be understood as an Oedipal conflict played out by an entire people: the son kills, devours, and internalizes the father to absorb his authority.[66] Alexis de Tocqueville had taken the first step toward this diagnosis a half century before Freud supplied the tools to make it plausible, remarking that "the King's subjects felt towards him both the natural love of children for their father and the awe properly due to God alone."[67] As king and surrogate for the ultimate Father, Louis XVI becomes the father who must be sacrificed for the people to absorb his authority. The people absorbing his authority are represented by a succession of tyrant-leaders: Danton, Robespierre, and finally Napoleon. Frye notes that at its most extreme, the kind of demonic parody the Revolution appears to suggest unites tyrant-leader and sacrificial victim in the same person.[68] Danton, Robespierre, and Napoleon are all sacrificed: the first two to the Revolution, the last to the Restoration.

If the poet never imitates life, life may (as Wilde not so puckishly proposed) imitate the poet. The tendency of life to imitate poetry attests to the formative power of archetypes. Whether romance is history refracted toward myth, as Scott believes, or myth refracted toward history, as Frye believes, makes no difference. Either way, romance cannot be divorced from history. Literature (as Graff sums up Frye's position) does not withdraw from the world. By crystallizing mythic archetypes, it explores possible approaches to comprehending the world.[69]

Paul de Man, surely one of those cognitive atheists against whom both Graff and Hirsch direct their fire, all but concedes literature's involvement in the world, at least by analogy, when he points out that the perfect forgetting of antecedents that Nietzsche hypothesizes as the ideal mental state for the man of action exceeds human grasp. To become conscious of one's modernity, whether as a revolutionary or as a writer, is to be aware of one's relation to a past. If we, as de Man insists, know that past only from texts, the facts comprising our texts were once tangible realities. "History," de Man acknowledges, "is not fiction."[70]

To render their history a fiction, that is, to cut themselves off from their own past, was, however, quite literally what the French attempted after the execution of Louis XVI—even to instituting a new calendar, to start with the Year One of the Revolution. And instituting a new calendar, both Walter Benjamin and Herbert Marcuse argue, satisfied an impulse evident in Parisians as early as July 1789, when they shot at clocks on their march to the Bastille. They had come to recognize in the concept of time as a steady, inexorable flow an instrument of tyranny, a message to the oppressed that the future lay beyond their capacity to change. Thus Benjamin rejects history devoted to "telling the sequence of events like the beads of a rosary," and calls on the Marxist historian to develop instead "a conception of the present as the 'time of the now' which is shot through with chips of Messianic time."[71]

To understand the present as shot through with chips of Messianic time is, ironically for a Marxist like Benjamin or Marcuse, to understand history as Blake and Coleridge had understood it: as the snake with its tail in its mouth. "In the true utopia," Marcuse declares, "time would not seem linear, as a perpetual line or rising curve, but cyclical, as the return in Nietzsche's idea of the 'perpetuity of pleasure.' "[72]

A hostile critic, either a fellow traveler of the old regime or an orthodox Marxist, would surely dismiss this contention as a wish-fulfillment fantasy. Tocqueville, whose attitude toward the Revolution remained ambivalent, and who saw the political order it produced largely as the old regime by another name, suggests just that: "Turning away from the real world around them, [Frenchmen] indulged in dreams of a far better one and ended up by living, spiritually, in the ideal world thought up by writers."[73] They sought to transform life into romance. That they had to fail is Tocqueville's thesis: "The peculiarities of our modern social system" he asserts in his foreword to *The Old Regime and the French Revolution* (1856), "are deeply rooted in the ancient soil of France."[74]

The emergence of modern historiography is deeply rooted, as Ranke and, later, Palmer remark, in the struggles that gave rise to France's social system.[75] Men found themselves compelled to reconcile two worlds: one, the world of the old regime, seemingly dead; the other, Orc-like, announcing its enraged right not merely to life but to power. In England, as elsewhere, this effort at reconciliation persistently manifests itself in reflections on the Revolution in France. For in asking what the French Revolution was about, measuring it against their own Glorious Revolution, Englishmen were asking what they and their society were about.

TWO

Through Forests of Eternal Death: Blake and Universal History

Seeing the Signs

In 1720, thirty-seven years before the magically Swedenborgian year of Blake's birth, Charles Daubuz distinguished history from prophecy, arguing that "an Historian sets out the matters he relates in proper Words, such as we express our Conceptions by, and therefore shews the full Extent of the Things acted, because his Words are adequate to our Notions: But a Prophecy is a Picture or Representation of the Events in Symbols; which being fetched from Objects visible to one View, or Cast of the Eye, rather represents the Events in Miniature, than full Proportion; giving us more to understand than what we see."[1] Showing the full extent of things acted in proper words suggests Blake's insistence, in the *Descriptive Catalogue of Pictures*, that the historian withhold reasons and opinions concerning acts, and focus instead on the acts themselves, which alone are history. Representing events in miniature, giving us more to understand than what we see, suggests the question Blake, in *The Marriage of Heaven and Hell* (1790), reports perceiving a Devil burn into the face of the abyss—"How do you know but ev'ry Bird that cuts the airy way, / Is an immense world of delight closed by your senses five?" (pl. 7; E, p. 35)—or the claim Blake, in *Milton*, makes in his own prophetic voice: "Every Time less than a pulsation of the artery / Is equal in its period and value to Six Thousand Years" (pl. 28: 62–63; E, p. 126).

The Devil's question anticipates the Fairy's mocking statement in *Europe* (1794) (included in only two of ten extant copies) that, of the "Five windows" lighting "the cavern'd Man," he may look through one, "And see small portions of the eternal world"; through another, "himself pass out . . . but he will not" (pl. iii: 1–5; E, pp. 58–59). It also anticipates the promise Blake, at the end of *America* (1793), foresees in the Revolution, that an apocalyptic day looms when the cavern will be itself consumed away: "the five gates were consum'd, and their bolts and hinges melted / And the fierce flames burnt round the heavens, and round the abodes of men" (pl. 16: 22–23; E, p. 56). Space as delineated by geometers' instruments, time as told by clocks, will dissolve, and man will emerge as bard or prophet, "who Present, Past, & Future sees."[2]

In his Prophecies, Blake epitomizes what C. A. Patrides, following a long tradition, calls universal history: the use of one nation's experience, or a particular event—here specifically the American and French Revolutions and the agony they cause Albion—as a microcosm (Daubuz's representation of the event in symbols) of the whole cosmic design.[3] Blake would have found a secular precedent for so treating the American Revolution in *Common Sense* (1776), where Paine declares America's cause to be *"in a great Measure the Cause of all Mankind."* Blake would also have found at least a mutedly chiliastic reading of this sign of the times in *Observations on the Importance of the American Revolution* (1784), where Reverend Richard Price, citing reason as well as tradition and revelation for authority, looks forward to "a more improved and happy state of human affairs [to] take place before the consummation of all things."[4]

Blake's view of history comes closest, however, to the views of those radical Protestant commentators on Scripture and society, cited by Patrides, who were convinced that the End predicted by John's vision on Patmos was almost at hand.[5] In a tract entitled *The Signs of the Times*, which appeared the same year Blake engraved *America*, James Bicheno speculates that the French Revolution may announce "the beginning of the fulfillment" of the prophecy "thy kingdom come." Four years later, in *The Probable Progress and Issue of the Commotions which Have Agitated Europe Since the French Revolution* (published by Joseph Johnson, whose authors included Blake and Paine), Bicheno adds that "Society appears as though in pangs for the birth of some new order of things; and those governments which have stood the blasts of ages, are, all at once, if not overthrown by the assaults of a new species of enthusiasm, yet tottering to their foundations before the breath of opinion; or rapidly sinking under the loads of debts and taxes, which, by wanton wars abroad, and boundless extravagance and corruption at home, they have brought upon themselves."[6]

Society suffering birth pangs recalls the condition of the daughter of Urthona in *America* after her ravishment by Orc, and the subsequent gestation and birth of the new, revolutionary Orc, who rises "a Wonder o'er the Atlantic sea" (pl. 4: 6; E, p. 53). Orc rising prefigures the all but hysterical response to revolutionary France voiced by Alexander Pirie in 1795: "It is a beast rising out of the *bottomless pit*, or vast abyss, as its politics are mischievous and deep as hell, and its actions works of the Devil." As Blake observes, from the contrary perspective of his *Marriage of Heaven and Hell*, "this history [of the struggle between God and Satan for control of the human world] has been adopted by both parties" (pl. 5; E, p. 34). In Pirie's words, "Satan has all along imitated the divine plan, in carrying on his opposition to the measures of divine government."[7]

The adoption of this history by both parties is rhetorically manifest in the work of Gerrard Winstanley, whose social radicalism often anticipates Blake's (and Paine's), but who, in his *New Law of Righteousness* (1648), proposes an interpretation of the story of Jacob and Esau that Blake a century and a half later was to reverse. For Winstanley, Esau repeats man's first disobedience, coalescing with the Adam who transgressed six thousand years ago to reappear in every man and woman:

> It is the first power that appears to act and rule in every man. It is the Lord *Esau* that stepped before *Iacob*, and got the birthright, by the Law of equity was more properly *Iacobs*.
>
> Though *Iacob*, who is the power and wisdom that made flesh did draw back, and give way, that the wisdom and power of flesh should possess the Kingdom, and rule first; till *Esau*, by delighting in unrighteousness, lost both birth-

right and blessing; and left both in the hand of *Iacob* the King, that rules in righteousness, that is to rise up next.[8]

Thus Winstanley, despite his political kinship with Blake, envisions Jacob—as Blake himself in *The Marriage of Heaven and Hell* was to envision Esau—coming from Edom to herald the New Jerusalem.

As Winstanley allegorizes the biblical story, "the Antichristian Captivitie is expiring. . . . Israel's Captivitie in the 70 years in *Babylon*, was but a type of this Antichristian slavery under the L. *Esau*, the powers of the flesh, that compasses mankind about with many straits and dangers, for acknowledging his Maker."[9] Although Blake would have utterly rejected Winstanley's orthodox conjunction of flesh with evil, he essentially adopts Winstanley's method of reading Scripture typologically, translating it into allegories that have contemporary relevance.[10]

As Winstanley discovers in the Babylonian captivity a type of Esau's reign over Jacob, the domination of man by the flesh, Blake discovers in the Israelites' bondage in Egypt a type of England's colonial rule in America, the domination of men by tyrants. In Blake's Prophecy, Washington foresees that "Albions fiery Prince" threatens to reduce the colonists to Israelite slaves, compelled to make bricks from straw:

> . . . a heavy iron chain
> Descends link by link from Albions cliffs across the sea to
>   bind
> Brothers and sons of America, till our faces pale and
>   yellow;
> Heads deprest, voices weak, eyes downcast, hands work-
>   bruis'd,
> Feet bleeding on the sultry sands, and the furrows of the
>   whip
> Descend to generations that in future times forget.
>
> (*Am.*, pl. 3: 7–12; E, p. 51)

Leslie Tannenbaum argues that Blake relates the colonists to the ten lost tribes, an idea suggested to him by the calculations of Francis Lee, in the popular *Dissertation on the Second Book of Esdras* (1752), that the place where their eighteen-month trek led them, "if we allow them to travel by very easy Marches at the rate of Eight or Ten Miles a Day, cannot have any where a Subsistence but in the *Atlantic* Ocean, with the Old or New Atlantis."[11] Tannenbaum's argument has merit, especially given the case he persuasively makes for recognizing in Esdras a model for *America,* and given the focus of Esdras on the return from the Babylonian Captivity, which its author treats as a second Exodus.

But the presence of plague as a shaping force in Blake's mytho-history suggests the first Exodus rather than the second. The patriots arrayed on America's shore to hear Washington's warning—Franklin, Paine, Warren, Gates, Hancock, and Green—reemerge (with a slight change: Allen and Lee replace Hancock and Green) to confront the armies of Albion's Angel after his arousal by Orc. Albion's Angel himself, sending forth his plagues to "blight . . . the tender corn" (*Am.,* pl. 14: 6; E, p. 55), appears to the Americans as the Angel of Death, burning "outstretched on wings of wrath cov'ring / The eastern sky, spreading his awful wings across the heavens" (*Am.,* pl. 13: 11–12; E, p. 55).

Dreadful Appearances

Tyranny as a malignant form hovering over its victims recurs in various guises in the illuminations of both *America* and *Europe*: as Urizen seated on a cloud with arms outstretched (*Am.,* pl. 10; B, p. 127), mirroring Orc, his arms similarly outstretched, enveloped in flames (*Am.,* pl. 12; B, p. 129); as an eagle, its wings spread, feeding on the body of a woman (*Am.,* pl. 12; B, p. 132); as a bat-winged Albion's Angel—Urizen by another name—wearing the Pope's triple tiara and enthroned above two bowing attendant angels, his brazen book open on his lap (*Eur.,* pl. 14; B, p. 217). A comparable apparition—"An aged form, white as snow, hov'ring in mist" (*FR* 7: 131; E, p. 288)—urges the suppression of the people in the dream recounted by the Archbishop in Blake's unengraved poem, *The French Revolution* (1791).

David Erdman traces the figure of Albion's Angel massing his plagues against America ultimately to John's vision (Revelation 15:1) of the "seven angels having the seven last plagues; for in them is filled up the wrath of God."[12] Bicheno, reading the "Commotions which Have Agitated Europe Since the French Revolution" as a sign of the coming new heaven and new earth, appeals to the same sacred text: "In the accomplishment of these awful, but in the end, glorious designs of God, by the execution of the seven last plagues, we may expect that there will be considerable correspondence of events with those of the trumpets which gave birth to the present kingdoms of Europe."[13]

Blake, however, derives his rationale for asserting the correspondence of events with biblical prophecy at least in part from Joshua Barnes's *History of Edward III* (1688), in which plague becomes—as it does in *America*—a judgment on England for its aggression against France.[14] As the Black Death is brought west across the Channel by Edward's army, the pestilence Albion's Angel marshals against America recoils east across the Atlantic upon his army, his people, himself. Indeed, Barnes suggests (if unconsciously) the association between plague and the king, as surrogate for the Angel of Death, when he describes Edward on July 11, 1346, six weeks before he defeated Philip VI at Crécy, taking "an high Hill near the Shore; from whence he made a Dreadfull Appearance over all the Country."[15] That Blake was struck by this image is clear from his appropriation of the episode of which it is a part—the knighting of Edward's son, the Black Prince—for the opening scene of his unfinished play, *King Edward the Third*, and from his elaboration of the image to ironic purpose in Edward's address to Liberty:

> I see thee hov'ring o'er my army, with
> Thy wide-stretch'd plumes; I see thee
> Lead them on to battle;
> I see thee blow thy golden trumpet, while
> Thy sons shout the strong shout of victory!
>
> (1.3: 204–9; E, p. 423)

The figure Edward sees hovering over the English army is stripped of its rhetorical disguise by Sir Walter Manny, who remarks of the battle into which his king seeks to lure the French that "death is terrible, tho' borne on angels' wings! / How ter-

rible then is the field of death" (1.5: 17–18; E, p. 427). Death
suspended above its victims, as though it were borne on angels'
wings, reappears at the beginning of *The French Revolution* to
epitomize the atmosphere of 1789: "The dead brood over Europe" (1: 1; E, p. 282). And the dead brooding over Europe
blend into the image, which occurs throughout both *The
French Revolution* and *America*, of the king ill with the disease
that afflicts his land. In the opening scene of *The French Revolution*, Louis XVI is portrayed as "Sick, sick: the Prince on his
couch" (1: 2). In *America*, George III is portrayed, after his assault on the colonists, as Albion's Guardian writhing "in torment," "Sick'ning," his limbs streaked with red (pl. 15: 6, 9, 1-
2; E, p. 55).

However aware Blake was of the tradition linking the health
of the king to that of the realm, he might have derived the political significance of this motif for his own time from Paine,
who asks in *Common Sense*, "Why is the constitution of England sickly?" and then answers, "because Monarchy have poisoned the Republic" (p. 16). Paine even shares one of his argumentative strategies with those Protestant commentators on
Scripture, whose faith he emphatically did not share, rooting
his indictment in 1 Samuel:

> In the early ages of the world, according to the scriptural
> chronology, there were no Kings; the consequence of which
> was, there were no wars; it is the pride of Kings which
> throws mankind into confusion. . . .
>
> Near three thousand years passed away, from the Mosaic
> account of the creation, till the Jews, under a national delusion, requested a King. 'Till then, their form of government
> (except in extraordinary cases where the Almighty interposed) was a kind of Republic, administered by a judge and
> the elders of the tribe. . . .
>
> Monarchy is ranked in scripture as one of the sins of the
> Jews, for which a curse in reserve is denounced against
> them. (pp. 8–9)

Paine subjects Samuel to a secular typological reading, in
which the Jews' desire for a king comprises a type of the Fall,
and, by inference, the revolutionary struggle against kingship

in America an adumbration of the restored Golden Age. He attempts to capture the dynamics of contemporary events, as Blake does, by incorporating them in a comprehensive mythohistory. Analytical and discursive rather than (like Blake's work) symbolical and pictorial, Paine's writing exemplifies the overlap that Hayden White detects between history and fiction. White characterizes the process of achieving historical understanding as tropological—rendering the unfamiliar familiar. The historian, he argues, evokes images, enabling the reader to comprehend events, to grasp their emotional impact.[16] The historian, that is, builds analogies, and thereby invites the reader to participate in the reconstruction of the history being narrated. Language works in historical narratives as language works in fiction. It encodes instructions for creating not the thing itself (inaccessible to both reader and writer) but a trope analogous to it. The reader contributes to the unfolding of historical narrative in a way comparable to the way in which, as Joseph Wittreich observes, he contributes to the unfolding of Blakean prophecy. He supplies the connections between vision (verbal or iconic) and commentary.[17]

Washington's exhortation to "Friends of America" (Am., pl. 3: 6; E, p. 51; B, pl. 5, p. 122) is illustrated by the nude figure of Orc in the upper left corner of the page, broken chains dangling from his wrists, his form paralleling in its muscular curves the sweep of the effulgent A of the heading "A Prophecy," as he ascends toward heaven. Like the A that metamorphoses into leaves and grain, the comparably ornate W of "Washington" blends into flames blown from a serpent-mouthed trump of a second nude figure, stretched in the air above the general's address. The flames run across the page, down the left margin, and beneath the speech, framing it on three sides. In the lower left corner are three more nude figures—a man, woman, and child, presumably members of a family—who, hand in hand, flee the conflagration.

Revolution merges into apocalypse, with all its consequences, destructive as well as creative. Though Orc's freedom is accompanied by the flight of birds and fecundity in nature, innocents are displaced and threatened by fire. Space, defined in the main pictorially, and time, defined in the main poetically, become space-time. The mind's eye of the reader becomes one with the mind's eye of the artist.[18]

Fire and Frost

The dimensions of Blakean mytho-history, described by visual and verbal media in concert, have an ambivalence comparable to that of Orc's rape of the daughter of Urthona. The flames that punctuate Washington's call to his countrymen when they confront "Albions wrathful Prince" (*Am.*, pl. 3: 14; E, p. 51) are reciprocated by the Prince himself in Urizenic form. Having been roused by Orc, and having rallied his armies, he looms on the shore as Orc's mirror image, a "Demon red, who burnt towards America" (*Am.*, pl. 12: 9; E, p. 54). He embodies a kinship between repression and revolution, reinforced by their parallel portraitures: Urizen's (B, pl. 10, p. 127) preceding a speech by Orc, Orc's (B, pl. 12, p. 129) following a speech by Urizen's surrogate.[19]

Similarly, Orc's rape, while expressing his defiance of Urizenic law, impregnating the daughter, even suffusing her womb with joy, nonetheless makes her the sort of tyrant that Urthona, who rivets Orc's "tenfold chains" (*Am.*, pl. 1: 12; E, p. 50), has been, and that Enitharmon, who forges her chain from social and religious mores, will be. "I have found thee," the daughter half exults, half laments over Orc, "and I will not let thee go" (*Am.*, pl. 2: 6; E, p. 51).

Refusing to let her lover go, binding him by an act of love, also describes Orc's embrace of the daughter: "Round the terrific loins he siez'd the panting struggling womb" (*Am.*, pl. 2: 3; E, p. 50). This tension between freedom and bondage, love and violence, is accentuated by an unfolding pattern of alliteration that interlocks images of rolling, riveting, rending, and surrounding. Despite Alicia Ostriker's *Vision and Verse in William Blake* (1965), too little attention has been paid to Blake's prosody in the Prophetic Books. Blake stresses the ironic consequences of the campaign by Albion's Guardian to destroy the Friends of America with pestilence not only through alliteration but through line break and plate division, isolating the king's own minions as chief victims of his attack: "The red

Fires rag'd! the plagues recoil'd! then rolld they back with Fury [pl. 15] On Albions Angels" (*Am.*, pl. 14: 20–pl. 15: 1; E, p. 55).

The image of plague "recoiling" on Albion's Angels configures the myriad serpents winding their way, poetically and pictorially, through *America*. This image is itself underscored (B, pl. 16, p. 133) by a tree root that becomes a serpent breathing flame. The verse passage to which it belongs is overhung (at the center of the page) by a branch sheltering a seated man, nude but for a cloth on his head. A phallic snake emerges from between his legs toward a robed, reclining youth whose elbows are propped on a stack of books, his hands clasped, perhaps in prayer. "Recoil'd," moreover, alliteratively calls up the Demon red, who threatens America with flame, "rejoicing in its terror" (*Am.*, pl. 12: 10; E, p. 54). Blake's syntax is brilliantly ambiguous. The Demon rejoices both in the terror it arouses in its victims and, apparently, in the terror aroused in it by the challenge issuing from America. Its joy recalls the joy felt by the daughter in the throes of ravishment. She hails Orc as the sacrificial god, who has "fallen to give me life in regions of dark death" (*Am.*, pl. 2: 9; E, p. 51) but who also, in impregnating her, has doomed her to those consequences of the Fall—the pain of childbirth and death: "O what limb rending pains I feel. thy fire and my frost / Mingle in howling pains, in furrows by thy lightnings rent; / This is eternal death; and this the torment long foretold" (*Am.*, pl. 2: 15–18).

Blake weaves into the daughter's outcry a complex of metaphors, to be repeated in his symbolic distillation of the struggle between England and America, both to establish the interdependence of Preludium and Prophecy, eternity and the productions of time, and to emphasize, again, the ambivalence with which he views the coming upheaval. Although the "furrows" the daughter suffers during Orc's assault sum up the labor she, like nature, must undergo to be rendered fertile, they evoke as well the "furrows" impressed by the taskmaster's whip on the backs of Israelite slaves (*Am.*, pl. 3: 11; E, p. 51), and promised to "Brothers and sons of America" (*Am.*, pl. 3: 9) should they yield to Albion's Guardian Prince. The daughter's howls echo the enchained Orc's howls of joy when she appears with his food (*Am.*, pl. 1: 19; E, p. 50), and the Israelites' howls of pain under the Egyptian lash. They sound in cacophony with the howls of the lions and wolves who, in both *America* and *Europe*, roam the "forests of eternal death" (*Eur.*, pl. 2: 6; E, p. 60) that constitute Blake's world.

The word in the daughter's lexicon that has the greatest symbolic reverberation in *America*, however, is *rend*. The furrows "rent" in her limbs and trunk by Orc's lightnings bespeak a power also manifest in his "rending" first of the caverns within which he is enchained (*Am.*, pl. 1: 18; E, p. 50) and then of the chains themselves (*Am.*, pl. 2: 2; E, p. 50). Blake's portrait of Orc breaking out of his earthly prison, his shoulders tensed, his head toward heaven (B, pl. 4, p. 121), foreshadows Boston's Angel "rending off his robe [in effect his "mental chains" (*Am.*, pl. 13: 3; E, p. 54)] and throwing down his scepter" (*Am.*, pl. 12: 1; E, p. 54), setting an example that is immediately followed by his twelve colleagues: "and all the thirteen Angels / Rent off their robes to the hungry wind, and threw their golden scepters / Down on the land of America" (*Am.*, pl. 12: 2–4).

George Quasha focuses on *rend* in his analysis of the implications of rape—what he calls the rending plots of *Visions of the Daughters of Albion* and *America*—arguing that the energy of Blake's poetics generates a kind of "torsion." Discovering in his dialectical verbs the minute particulars or "discrete terminological units" of Blake's verse, Quasha shows how Blake forces words to turn contrary ways—to imply themselves and their opposites in an expanding and reciprocal rhythm that embraces the furthest bounds of symbolic meaning. (Orc and Urizen, Quasha observes, embody inimical yet interdependent human possibilities.)

Each crucial word contains a cosmos.[20] As Blake learned from those Proverbs he collected while walking in Hell, "One thought fills immensity" (*MHH*, pl. 8: 36; E, p. 36). To the Poet-Prophet, any given moment may become a paradigm of universal history[21]—which is why *America* and *Europe*, as well as *The French Revolution* and *The Four Zoas*, can simultaneously dramatize the vicious politics of Blake's world and the visionary politics of eternity. Though Harold Bloom has, I think, convincingly dismissed the once fashionable view that Blake was steeped in the occult, these early Prophetic Books seem to posit a cosmos that functions according to the Hermetic axiom "as above, so below." The "warlike men, who rise in silent night" (*Am.*, pl. 3: 3; E, p. 51) on America's shore parallel Orc when he breaks out of his prison and then rises over the Atlantic. The thirteen Angels who descend, like Milton's fallen angels, "headlong from out their heav'nly heights" (*Am.*, pl. 12: 5; E, p. 54) gather around Bernard as "the thirteen Governors that England sent" (*Am.*, pl. 13: 1; E, p. 54). Bernard himself declares his independence in a speech that echoes not only Washington's rallying call to his fellow patriots but Orc's proclamation of a new heaven and a new earth. "What God is he," Bernard demands in his repudiation of the old regime, "writes laws of peace, and clothes him in a tempest / What pitying Angel lusts for tears, and fans himself with sighs / What crawling villain preaches abstinence and wraps himself / In fat of lambs? no more I follow, no more obedience pay" (*Am.*, pl. 11: 12–15; E, p. 54).

Spirits of Fire

Like Washington, Franklin, Paine, Warren, Gates, Hancock, Green, Allen, and Lee, Bernard and his colleagues exist in both political and millennial—revolutionary and apocalyptic—realms. The dimensions of the cosmos in which America and France struggle for freedom from ancient tyranny are symbiotically knit. In *The French Revolution* the Bastille's dens shake; the prisoners look up, shout, and laugh; "and a light walks round the dark towers. / For [because] the Commons convene in the Hall of the Nation" (*FR* 3: 52–54; E, p. 285). In *America* Albion's Angel's "punishing Demons" can neither blight the soil, nor ring cities and castles with isolating walls and moats, nor "bring the stubbed oak to overgrow the hills. / For [because] terrible men stand on the shores" (*Am.*, pl. 9: 8–9; E, p. 53).

The revolutionary heroes of France and America affect spiritual as well as temporal reality. Sieyès's demand that the king's troops withdraw from Paris—"Then hear the first voice of the morning: 'Depart, O clouds of night, and no more / Return . . .'" (*FR* 12: 238–39; E, p. 293)—anticipates Orc's warning to Albion's Angel: "The times are ended; shadows pass the morning gins to break" (*Am.*, pl. 8: 2; E, p. 52). That Blake translates his human spokesmen for this new morning into characters in his visionary world, yet retains their human identities, gives to mortal beings mythic stature, makes of them giant forms. They confront tyranny not only as the names that fill the histories of Blake's age but as types of biblical prophets. The patriots who stand on America's shore, "their foreheads reard toward the east" (*Am.*, pl. 9: 11; E, p. 53), recall Ezekiel, assured by God that "I have made thy face strong against their faces, and thy forehead strong against their foreheads" (3:8). The Commons convening in the Hall of the Nation, "like spirits of fire in the beautiful / Porches of the Sun, to plant beauty in the desert craving abyss" (*FR* 4: 54–55; E, p. 285), participate in the power that plants "Roses . . . where thorns grow" (*MHH*, pl. 2: 5; E, p. 33). They redeem the wasteland where, as Isaiah (34:13) observes (in a chapter cited in *The Marriage of Heaven and Hell*), only thorns, nettles, and brambles have flourished.

And Orc, announcing his new birth, in a speech Erdman labels the Declaration of Independence recast as a sequence of images, is also recasting Ezekiel's apocalyptic vision in the valley of dry bones.[22] What Erdman interprets as Orc's evocation of life (the first component in Jefferson's trinity of inalienable rights)—"The morning comes. . . . The grave is burst. . . . The bones of death, the cov'ring clay, the sinews shrunk and dry'd. / Reviving shake, inspiring move, breathing, awakening!" (*Am.*, pl. 6: 1–4; E, p. 52)—echoes Ezekiel's account of the miracle God shows him:

> So I prophesied as I was commanded: and as I prophesied, there was a noise, and behold a shaking, and the bones came together, bone to his bone. And when I beheld, lo, the sinews and the flesh came up upon them, and the skin covered them above: but there was no breath in them. Then said he unto me, Prophesy unto the wind . . . and say unto the wind, Thus saith the Lord God; Come from the four winds, O breath, and breathe upon these slain, that they may live. So I prophesied as he commanded me, and the breath came into them, and they lived, and stood up upon their feet, an exceeding great army. (37:7–10)

I do not mean that the voice behind Orc's voice is solely Ezekiel's; that would be to oversimplify Orc's utterance. His image of "the linen wrapped up" (*Am.*, pl. 6: 2) is an allusion to the discovery of Christ's vacant tomb; as Tannenbaum points out, the watchmen, whom Orc reports leaving their stations (pl. 6: 1), have prototypes in both Ezekiel and Isaiah.[23] Orc's Declaration gathers up various of the apocalyptic moments that climax scriptural narrative. It is a prophecy within the Prophecy. Headed (B, pl. 8, p. 125) by a portrait of Orc seated on a grave from which the skull protrudes, his face looking heavenward, his hands braced against the mound as if he is about to rise, the Declaration is followed (B, pl. 9, p. 126) by an idyllic scene of two children and a ram asleep under a tree populated with exotic birds. This scene, suggestive of the pastoral paradise regained in Night IX of *The Four Zoas*, envisions the fulfillment of Orc's promise that, with empire no more, "the Lion and Wolf shall cease" (pl. 6: 15; E, p. 52).

That his promise remains unfulfilled is established textually by Albion's Angel's reply to Orc, set above the scene of the ram and children and framed on three sides by the earth on which they lie and the trunk and branches of the tree (perhaps a weeping willow). In *America*, Blake reverses the sequence from Isaiah that he cites in *The Marriage of Heaven and Hell*, in which Isaiah's nightmare of the land as a wilderness possessed by the cormorant and the bittern, the owl and the raven (34:10–11), yields to a vision of the land redeemed: "No lion shall be there, nor any ravenous beast ... go up thereon" (35:9). This reversal may explain the rage that provokes the Bard (in the first and last two etched copies of *America*) to smash his harp and pass "down the vales of Kent in sick and drear lamentings" (*Am.*, pl. 2: 21; E, p. 51). Blake seems, as Nelson Hilton suggests, to concede (momentarily) the inability of the Poet-Prophet to shape events.[24] If "A Song of Liberty" in *The Marriage* ends with the same proclamation that ends Orc's Song of Liberty in *America*—"Empire is no more! and now the lion and wolf shall cease" (*MHH* 25; E, p. 44)—the wrath of Albion's Angel in defying Orc is likened to "the Eternal Lions howl" (*Am.*, pl. 7: 2; E, p. 52); the alarm he sounds to rally his armies and his thirteen Angels opens and closes with the warning, "Loud howls the eternal Wolf! the eternal Lion lashes his tail!" (*Am.*, pl. 9: 2, 27; E, p. 53).

Looking toward America, spying Orc risen over the Atlantic, Albion's Angel sees the very evil that Orc sees when he looks toward England and spies Albion's Angel burning "beside the

Stone of Night" (*Am.*, pl. 7: 2; E, p. 52). Where Orc discerns in his new birth the shadows passing and morning beginning to break, Albion's Angel discerns "America . . . darkened" (*Am.*, pl. 9: 3; E, p. 53). Where Orc discerns shrunken, dried sinews shaking, reviving, Albion's Angel discerns his Demons "before their caverns deep like skins dry'd in the wind" (*Am.*, pl. 9: 4; E, p. 53).

The symbolic wind that shrivels the Demons before their (tomblike) caverns blows to the opposite effect from the four winds on which Ezekiel calls in his prophecy, or, their counterpart, the breath that, in the prophecies of both Ezekiel and Orc, inspires the dead to rise. Blake is using a device to be used again by the poet of a great, historically rooted poem, Browning, who in *The Ring and the Book* (1868–1869), also dramatizes conflict by having antagonists express themselves through the same metaphors, but to contrary purposes.

Henry James praised *The Ring and the Book* as the first point-of-view novel. But Blake, as several recent critics have stressed, had long since mastered the technique of opposing different perspectives within a narrative, albeit in a way that departs from conventional plotting much more radically than Browning does. Erdman characterizes *America* as an "acting version of a mural apocalypse," an entire historical cycle viewed from its point of completion in 1793. Tannenbaum, building on Erdman's interpretation, finds the poem to be a "visualizable picture of shifting perspectives."[25] No one would dispute the importance of this insight for understanding Blake's prophetic form. W.J.T. Mitchell devotes much of his study of the illuminated books to exploring Blake's use of poetry and pictures in tandem to break down our habitual, compartmentalized ideas of time and space.[26]

The trouble is that only after repeated readings (and maybe not even then) can one experience *America*, or *Europe*, as a mural. Mitchell himself, looking back on his immersion in Blake's work, acknowledges his inability to address the poem and pictures simultaneously. That Blake recognized the problem he was posing for the reader is reflected in the care with which he specifies the time elapsed between the political events that underlie *America* and *Europe*—"Angels and weak men twelve years should govern o'er the strong: / And then their end should come, when France receiv'd the Demons light" (*Am.*, pl. 16: 14–15; E, p. 56)—and with which he establishes the Preludium to *Europe* as a continuation of the Preludium to *America*: "The nameless shadowy female rose from out the breast of Orc" (*Eur.*, pl. 1: 1; E, p. 59). Blake is trying to help the reader "see a World in a Grain of Sand / And heaven in a Wild Flower / Hold Infinity in the palm of [his] hand / And Eternity in an hour" ("Auguries of Innocence," E, p. 481).

Continued Hieroglyphics

To see the world in a grain of sand, heaven in a wild flower is precisely what the Fairy makes possible for the Blakean questioner in *Europe*. The Fairy responds to the questioner's crucial question, "what is the material world and is it dead?" (pl. iii: 13; E, p. 59), by offering to "write a book [*Europe* itself] on leaves of flowers" (pl. iii: 14), and by showing his companion, within the wild flowers he gathers, "each eternal flower" (pl. iii: 20).

The Preludia and Prophecies of *America* and *Europe* are linked as are those eternal and temporal flowers. Orc's rape of the daughter, her lament to Enitharmon, occupy a kind of timeless moment—what Frank Kermode (borrowing the concept from Aquinas) might label an *aevum*: a realm whose inhabitants, immortal and therefore free of time and succession, seem nonetheless to act in time and succession.[27] This timeless moment correlates with, and in a sense causes the history initiated by the call to arms of America's patriots, and extending first to 1781, when the English were forced to concede that the end of their hegemony in America had come, then (Blake's twelve-year hiatus, during which the weak continue to rule the strong) to 1793, when the hapless Louis XVI discovered that his end had come. And history, as delimited by the American and French Revolutions, embodies in microcosm universal history: beginning with the Fall; evolving through the Incarnation, Passion, and Resurrection of Christ, the consolidation of error in Bacon and Newton, and the upheavals of Blake's own day; looking ultimately to a millennium, which is also a return to a Golden Age.[28]

Blake reads history as George Stanley Faber, in *The Sacred*

Calendar of Prophecy (1828), reads Revelation: as a "continued hieroglyphic."[29] This hieroglyphic embraces pagan as well as Christian history, for "All Religions are one": "The Religions of all Nations are derived from each Nation's different reception of the Poetic Genius which is every where call'd the Spirit of Prophecy" (E, p. 2). The symbolism of *America* and *Europe* thus stems from apocalyptic narratives not only in Scripture but in the Eddas, as Blake encountered them in Paul Henri Mallet's *Northern Antiquities* (1752). "There will come ... says the Edda [in Bishop Percy's translation of Mallet] ... a barbarous age," succeeded by "a desolating winter" the length of three in which "the snow shall fall from the four corners of the world, the winds shall blow with fury, the whole earth shall be bound in ice." In *America* this winter is personified by Urizen who, "clothed in ... trembling shuddering cold," pours forth "stored snows" (pl. 16: 8–9; E, p. 56) to quench Orc's flames. The catastrophe it portends (as foretold by the Edda) will be brought on by monsters who, Orc-like, break their chains and escape, and by a "great Dragon," like the dragon form of Albion's wrathful Prince (*Am.*, pl. 3: 14–15; E, p. 51), who "shall roll himself in the ocean, and with his motions the earth shall be overflowed."[30]

The earth overflowed, its archetypal occurrence recorded in Genesis, is a recurrent symbol of the Fall and the onset of history, not only in Blake but in a number of earlier biblical commentators in the seventeenth and eighteenth centuries. Jacob Bryant, whose *A New System, or, an Analysis of Ancient Mythology* (1774) Blake illustrated, allegorizes "the stay in the ark ... [as] a state of death, and of regeneration. The passage to life was through the door of the Ark. ... Through this the Patriarch made his descent: and at this point was the commencement of time." Bryant cites in support of his exegesis the construction he claims the Egyptians put on the Flood and its aftermath—to be writ large in Blake's Prophetic Books—that "This renewal of life was ... esteemed a second childhood. They accordingly in their hieroglyphics described him as a boy; whom they ... called Orcus."[31]

Blake in his *Descriptive Catalogue of Pictures* enlists Bryant to support his insistence that "the antiquities of every Nation under Heaven, is no less sacred than that of the Jews" (E, p.

534). All Floods, like all religions, are one. America and England are divided because the ocean has "barr'd out" the Atlantean hills (*Am.*, pl. 10: 6; E, p. 54). Had not the Americans rushed together against the depredations of Albion's Angel, America would itself be "o'erwhelm'd by the Atlantic, / And Earth had lost another portion of the infinite" (*Am.*, pl. 14: 17–18; E, p. 55). This whole fallen history—earth, portion by portion, bereft of the infinite—dates from the moment when "the five senses whelm'd / In deluge o'er the earth-born man; then turn'd the fluxile eyes / Into stationary orbs, concentrating all things" (*Eur.*, pl. 10: 10–12; E, p. 62).

Blake understands the Flood, much as Bryant does, as a trope for the malignant force that blinds man to the infinite in everything, to the deity in his own breast, and thus consigns him to forests of night. The imagery in *Europe* by which this understanding is conveyed—eyes fixed, "The ever-varying spiral ascents to the heavens of heavens . . . bended downward; and the nostrils golden gates shut / Turn'd outward, barr'd and petrified against the infinite" (pl. 10: 13–15)—prefigures the Blakean Geneses in *The [First] Book of Urizen* and Night IV of *The Four Zoas*. *Europe* contains in Enitharmon's dream a capsule history not merely of the eighteen hundred years following the birth of Christ but of the postlapsarian universe:

Thought chang'd the infinite to a serpent; that which pitieth:
To a devouring flame; and man fled from its face and hid
In forests of night; then all the eternal forests were divided
Into earths rolling in circles of space, that like an ocean rush'd
And overwhelm'd all except this finite wall of flesh.
Then was the serpent temple form'd, image of infinite
Shut up in finite revolutions, and man became an Angel;
Heaven a mighty circle turning; God a tyrant crown'd.

(pl. 10: 16–23)

The serpentine imagery pervading Blake's chronicle is reinforced by the picture of a serpent breathing fire (B, pl. 13, p. 216) that loops up the left side of the page and symbolizes the dilemma of history. When Albion's Angel, in *America*, confronts Orc across the Atlantic, he perceives his adversary as Ouroboric: "Eternal Viper self-renew'd" (pl. 9: 15; E, p. 53). Orc as self-renewed viper is a sign Albion's Angel supposes he can read: "now the times are return'd upon thee" (pl. 9: 19). Whereas Orc—born the rebellious son of Los and Enitharmon and reborn "a Human fire fierce glowing" (Am., pl. 4: 8; E, p. 52), the revolutionary son resulting from his own rape of the daughter of Urthona—perceives in his self-renewal a forecast of the end of time.

In Forests of Night

Whether Orc or Albion's Angel reads this sign of the times correctly is determined by those "immortal demons of futurity" (Eur., pl. 9: 10; E, p. 61) who rush down on the king and his bands as they flee the plagues recoiling against them from America. Is England's defeat by the colonists a prelude to the apocalypse, or is it one more turn in the endlessly turning wheel of history? Is the serpent that bends horizontally across the top of the page recounting the king's retreat to "his ancient temple serpent-form'd" (pl. 10: 11; E, p. 62. B, pl. 13, p. 216) to resume its upward progress, or to curl back until it bites its own tail? Is history in Blake's vision spiral or cyclical?

Spiral is the answer proposed by Lawrence Lipking, who argues that, for Blake in *The Marriage of Heaven and Hell*, wisdom is cumulative, fed by the inspiration of each prophet successively.[32] Spiral is also the answer Blake might have gleaned from Milton, who wrote his *History of Britain* (1670)—scenes of which Blake sketched—in the faith that his own troubled time might learn from a similarly troubled past:

> This third Book . . . may deserve attention more than common, and repay it with like benefits to those who can judiciously read: considering especially that the late civil broils had cast us into a condition not much unlike to what the *Britans* then were in, when the imperial jurisdiction departing hence left them to the sway of their own Councils; which times by comparing seriously with this intereign, we may be able from two such remarkable turns of State, producing like events among us, to raise a knowledge of ourselves both great and weighty. . . .[33]

And spiral seems finally to be the view Blake proposes in *America*, which ends with the five gates that shut man off from the infinite at last consumed.

Europe, however, pulls back from this view; it unfolds not to consummation but to conflict: Los calling "his sons to the strife of blood" (pl. 15: 11; E, p. 65). History is suspended in the "middest"; no end of times looms beyond its close. The serpent temple to which the king retreats incorporates the apparent paradox that Bard and daughter, as a result of their different torments, perceive in history. The daughter, vexed by the grasp of futurity, the vision of the infant Christ, to which Orc's embrace has brought her, asks "who shall bind the infinite with an eternal band? / To compass it with swaddling bands?" (pl. 2: 13–14; E, p. 60)—how, that is, can the infinite be "shut up in finite revolutions"?

"Revolutions" and "compass" are characteristic Blakean puns that denote the seemingly endless cycles of fallen man, with the upheavals triggering them, and re-evoke the figure on the frontispiece (B, pl. 1, p. 204): Urizen-Newton leaning out of his celestial sphere, a compass extended from his fingers to circumscribe space. He in effect encompasses the daughter's rape by Orc and the American and French Revolutions, the history of man from Creation onward, with which the daughter's rape is interwoven. Enitharmon falling asleep "in middle of her nightly song" (*Eur.*, pl. 9: 4; E, p. 61), dreaming her eighteen-hundred-year dream of Christian civilization, waking to Newton's "enormous blast" (pl. 13: 5; E, p. 63) upon the trump, "nor knew that she had slept" (pl. 13: 9; E, p. 64), then continuing her song as if those eighteen hundred years "had not been" (pl. 13: 11), seems to assert the inevitability of man's entrapment in the cycles of history. The narrative moves from present, to past, to present, offering no prospect (beyond the sound of the trump) of a millennial future. Enitharmon is not resurrected, merely reawakened.

The world to which she reawakens is still the world in which she has fallen asleep. The sons and daughters on whom she calls when she resumes her song comprise a kind of pageant (like Spenser's in *The Faerie Queene*), summing up the nature of that world: "Ethinthus, queen of waters . . . [shining] in the sky" (*Eur.*, pl. 14: 1; E, p. 64), is comparable to the moon, controller of tides, embodiment of cyclical change, of mutability itself; Manathu-Vorcyon "flaming in my halls" (pl. 14: 6) suggests the hearthfire, the inverse of Orc's consuming fire, and a "soft delusion" (pl. 14: 8), image of social hypocrisy; Leutha, "my lureing bird of eden! . . . silent love" (pl. 14: 9), her name combining *lure* and *lewd*, epitomizes repressed sexuality, "Sweet smiling pestilence" (pl. 14: 12); and she is followed by the victims of repression, male and female: Antamon, who would leave his mother, but who floats alone "upon the bosomed air: / With lineaments of gratified desire," his love sought by "the seven churches of Leutha" (pl. 14: 16–20); Oothoon, who would "give up womans secrecy" (pl. 14: 22); and her thwarted lover, Theotormon, theology-tormented man, "robb'd of joy," whose tears flow "down the steps of my crystal house" (pl. 14: 24–25).

Enitharmon's roll of sons and daughters concludes with Sotha and Thiralatha, to whom she appeals to "please the horrent fiend with your melodious songs" (pl. 14: 27), soothe him out of his revolutionary rage; and with Orc himself, whom she enjoins to "smile . . . and give our mountains joy of thy red light" (pl. 14: 29–30; E, p. 65). She is, she supposes, establishing her own peace. The "enormous revelry" her children enjoy is felt by "nature . . . thro' all her pores" (pl. 14: 34), like the distinctly more solemn joy Milton records as greeting Christ's Nativity. No one, I take it, need any longer be shown that the first voice heard in *Europe* after the nameless shadowy female falls silent—"The deep of winter came; / What time the secret child, / Descended thro' the orient gates of the eternal day: / War ceased, and all the troops like shadows fled to their abodes" (pl. 3: 1–4; E, p. 60)—parodies the opening stanzas of Milton's Hymn in "On the Morning of Christ's Nativity." As viewed by Blake in 1794, the respite from war that Milton depicts the Divine Birth bringing has been all too brief. The secret child who descends into Europe is Orc as well as Christ. When the revelries of Enitharmon's children end, and they return to their stations, Orc's fury envelops France in the red not of joy but of "wheels dropping with blood" (*Eur.*, pl. 15: 5; E, p. 65).

Blake inverts the symbolism Milton attaches to his new morning. While the "heav'n-born child" of the "Nativity" descends to a "happy morn" of peace on earth, the secret child of Blake's parodic nativity perceives "morning in the east" (pl. 14: 37; E, p. 65) as a signal to set Europe aflame. That Orc shoots from Enitharmon's heights at dawn, as Christ descends through the orient gates at dawn, identifies these events as one.

Europe constitutes a suspended moment, which is partly why Blake casts Enitharmon's recapitulation of Christian history as a dream. The narrative that follows the Preludium is quite precisely "A Prophecy": the epiphany of the Poet-Prophet, situated at a particular point in time, comprehending his age in the context of its past, and reading in the exfoliation of the present from the past the probable future.

Michael Tolley discerns in the poem's ending—dominated by furious Orc and illustrated (B, pl. 18, p. 221) by the picture of a nude man carrying a woman over his right shoulder and pulling a child with his left hand, all fleeing Orc's fire—an image of Armageddon.[34] Tolley's proposal is consistent with Blake's practice elsewhere. Indeed, the trio that portrays one consequence of Orc's descent into France's vineyards recalls the trio portrayed fleeing the flames in *America*. In *Europe*, however, Orc's fire does not, as it does in *America*, purge the senses to reveal the infinite that was hid. *Europe* ends with a promise of more slaughter: "The Lions lash their wrathful tails! / The Tigers couch upon the prey and suck the ruddy tide" (pl. 15: 6–7; E, p. 65).

These images of bestiality reinforce the daughter's lament to Enitharmon that impregnation by Orc causes her to "bring forth howling terrors, all devouring fiery kings. / Devouring and devoured roaming on dark and desolate mountains / In forests of eternal death" (*Eur.*, pl. 2: 4–6; E, pp. 59–60). Blake's Europe epitomizes a cosmos seemingly not providential but (to use a word that, if anachronic, is nonetheless irresistible) Darwinian. The daughter's distress over her progeny is partly framed by an illumination that runs along the bottom and up the right side of the page (B, pl. 5, p. 208) and shows a nude man throttling two other men, with a third in flight. In a world of devouring and devoured only the fittest survive. The preceding page (B, pl. 4, p. 207), containing the daughter's expression of despair over her own state, shows in what is actually the headpiece to the Preludium a pilgrim (Bunyan's Christian?) with a pack on his back, his face turned away from a cave, at the mouth of which crouches a man with a knife in his left (sinister) hand. Fitness consists in cruelty and cunning as well as strength. No wonder the title "A Prophecy" on the page following the daughter's lament (B, pl. 6, p. 209) is underscored by the picture of a winged female (the daughter herself? Eni-tharmon weeping at the departure of her children for war?) holding her head in her hands.

The Poet-Prophet's vision of Europe as a forest of predators informs the poem. Blake had, in both *The French Revolution* and *America*, already caught this element of nature red in tooth and claw in the struggle between liberty and tyranny. "Sudden seiz'd with howlings, despair, and black night" through contemplating the Commons convened in the Hall of the Nation, the Governor of the Bastille stalks "like a lion from tower / To tower" (*FR* 2: 21–22; E, p. 283). In the message of the divine visitant to his dreams the Archbishop hears the kind of rhetoric Blake uses to diminish the Governor used against the revolutionaries themselves, "a godless race / Descending to beasts" (*FR* 8: 138–39; E, pp. 288–89).

The Archbishop's orthodox Christian universe is ordered according to the Great Chain of Being, which, as Arthur O. Lovejoy notes, and as Blake knew well, necessarily serves the status quo. As flawed a creature as man can hardly gain political wisdom or attain virtue; to try is to aspire, Satan-like, to a place above that appointed for him on the Great Chain, and so, ironically, to fall instead of rise.[35] Sieyès, radical churchman and Blakean counterpart to the Archbishop, turns this doctrine upside down, prophesying a future in which men shall become as gods and the priest bless rather than curse, "That the wild raging millions, that wander in forests, and howl in the law blasted wastes, / Strength madden'd with slavery, honesty, bound in the dens of superstition, / May sing in the village, and shout in the harvest, and woo in pleasant gardens, / Their once savage loves, now beaming with knowledge, with gentle awe adorned" (*FR* 12: 227–30; E, pp. 292–93).

Like the Archbishop, Sieyès purports to speak words of a voice within his voice. Blake poses them, as he poses Orc and Albion's Angel in *America*, to mirror each other. In *The French Revolution*, if not quite in *America*, the world in which the lion and wolf shall cease appears at hand. On command of the nation's Assembly the army withdraws ten miles from Paris, and the resurrection, envisioned by Ezekiel in the valley of dry bones and by John on Patmos, happens:

. . . the bottoms of the world were open'd, and the graves of
arch-angels unseal'd;
The enormous dead, lift up their pale fires and look over the
rocky cliffs.

(16: 301–2; E, p. 296)

Even in *America*, though the colonists' refusal of Albion's Angel's "loud alarm" (pl. 10: 4; E, p. 54) cannot dissuade him from
bringing his plagues down on them, the conflagration Orc ignites in returning the angelic weapon against its sender opens
the doors of marriage, drives the repressive priesthood "into
reptile coverts" (pl. 15: 20; E, p. 56), and draws man toward the
infinite.

The plague that infects Albion's Angel and his bands also infects the soul of the Governor of the Bastille, "panting over the
prisoners like a wolf gorg'd" (*FR* 2: 26; E, p. 283). His wolfish-
ness distinguishes him from the prisoners who, confined to
"dens," seem degraded but retain the humanity their jailer has
lost—an irony Blake's prosody emphasizes. The image of the
Governor as gorged wolf, stressed by the halt to which it brings
the line, is balanced by the counterstatement, "the den nam'd
Honor held a man" (2: 26), equally stressed by its place at the
line's end. And this contrast is accentuated by the repetition of
"man" (or "woman") characterizing each of the unfortunates
chained in the Bastille.

By the time Blake engraved *Europe*, three years after he wrote
The French Revolution, the distinction had blurred. Blake has
reverted to a judgment on the conduct of fallen man strikingly
like Winstanley's in *The New Law of Righteousness*: "In times
past and times present, the branches of man-kind have acted
like the beast or swine; and though they have called one another, men and women, yet they have been but the shadows of
men and women" (p. 157). "Shadows of men in fleeting bands
upon the winds: / Divid[ing] the heavens of Europe" (*Eur.*, pl.
9: 6–7; E, p. 61) are what Enitharmon dreams of; and this division produces the society—which in her dreams she laughs to
see—that makes of "every house a den," thus binding every
man (pl. 12: 25–26; E, p. 63). Europe has itself become a Bastille.

The jailers Enitharmon appoints to govern it, "that Woman,

lovely Woman! may have dominion" (pl. 6: 3; E, p. 61), are Rintrah, whom she enjoins to "raise thy fury from thy forests black"; Palamabron, "skipping upon the mountains"; and Elynittria, "silver bowed queen" (pl. 8: 2–4; E, p. 61), who embodies secular and religious authority, and (in Blake's moral scheme) a perverse virginity akin to Diana's. Tannenbaum suggests that Rintrah and Palamabron recall Apollo and Hammon—(like Elynittria-Diana) pagan gods, banished, in Milton's version of the Nativity, by the incarnate Christ. Palamabron, "horned priest," suggests Hammon as a ram;[36] Enitharmon invokes Rintrah as a lion. Both reflect man acting like a beast.

Rintrah also reflects man acting as chivalric champion. Citing the headpiece to Enitharmon's claim of hegemony (B, pl. 8, p. 211)—Rintrah equipped with sword and armor, posed before two winged females, presumably the queens of France and England spiritualized—Erdman identifies him in *Europe*'s historical allegory as Pitt, whom Paine (mocking Burke's complaint that to dethrone Marie Antoinette was to destroy chivalry) pronounced the knight-errant of an order for which Burke sounded the trumpet. Blake perceives in chivalry a symptom of repressed sexual energy sublimated by war.[37]

Sexuality thus emerges in a perverse way as a major dynamic of history. Blake implies as much in Enitharmon's dream-chronicle of fallen man, evoking the scene in *Paradise Lost* where Adam and Eve, having discovered lust and shame, try to hide from God. In *Europe* (pl. 10: 16–18; E, p. 62), this scene is recast as man fleeing from the face of the infinite changed into a serpent, a devouring flame, to hide in forests of night. The temple to which the king, after his defeat in America, repairs with his band—that is, England, Europe, the world—has been transformed by the process that is human history into a spoiled Eden, depicted as those forests of night or, in the daughter's grimmer formulation, of eternal death. The southern porch, at which the ancient Guardian arrives, is "planted thick with trees of blackest leaf" (pl. 10: 25).

That his retreat takes him south indicates the 180-degree twist given the world, and man, by the Fall. The Stone of Night, platform for the Guardian's rise from his discomfiture, becomes in Blake's helpful explication of his own allegory an "image of that sweet south, / Once open to the heavens and elevated on the human neck, / Now overgrown with hair and covered with a stony roof" (pl. 10: 27–29). Like the ravished daughter, her "roots ... brandished in the heavens ... [her] fruits in earth beneath" (pl. 1: 8; E, p. 59), man has been cut off from the heavens and flipped on his head.

> Downward 'tis sunk beneath th' attractive north, that round the feet
> A raging whirlpool draws the dizzy enquirer to his grave:
>
> (pl. 10: 30–31)

"What Is the Material World and Is It Dead?"

The enquirer in *Europe* is made dizzy in part by inquiry itself. Thought, the onset of self-consciousness, is specified as the force that changes the infinite to a serpent. Self-consciousness, however, is an inevitable outgrowth of the Fall. "Enquirer" also describes the Blakean persona asking the Fairy, "What is the material world and is it dead?" (pl. iii: 13; E, p. 59). His questions sum up the crisis in European culture precipitated, to Blake's mind, by Newton. M. H. Abrams has pointed out that post-Restoration criticism, in its demand that art imitate nature, created the vexing problem of how to reconcile such a restrictive realism with myth, magic, fairy tale, even metaphors that violated observed fact.[38] Christopher Hill, in examining the permutation on this intellectual stance supposedly fathered by Newton, argues that the dismissal of nonmechanistic theories of nature (of a world in which God and Satan, magicians, witches, fairies, and the stars influence events) had ideological roots, and that these powers, usually connected with one form of radicalism or another, were perceived, especially after 1660, to threaten social order.[39]

Abraham Cowley takes essentially this position in his ode, "To the Royal Society" (1667), which Blake, had he read it, must have loathed. To Cowley, the struggle between science and superstition, for guardianship over "Philosophy the great and only Heir / Of all that Human Knowledge which has bin / Unforfeited by Mans rebellious Sin" (1–3), is a struggle for authority—the hero of which in Cowley's poem is to reappear as a villain in Blake's:

> *Bacon* at last, a mighty Man, arose
> Whom a wise King and Nature chose
> Lord Chancellour of both their Lawes
> And boldly undertook the injur'd Pupils cause.
>
> (37–40)

That Blake had indeed read and, with asperity, remembered these lines is strongly suggested by the entrance of Newton to seize and blow the trump in *Europe*—"A mighty Spirit leaped from the land of Albion" (pl. 13: 4; E, p. 63)—and by one of those mutedly savage comments Blake directed at Bacon's *Essays Moral, Economical, and Political*: "The Prince of darkness is a Gentleman and not a Man he is a Lord Chancellor" (E, p. 612). Newton seizing and blowing the trump thus represents Bacon as well as himself. Like Bacon, interpreting the laws of both king and nature, Newton consolidates error in physical, metaphysical, and political realms. Together, Newton and Bacon are what Erdman declares Newton alone (in *Europe*) to be: ideological founders of the anti-Jacobin universe. They formulate the ideology that pronounces the material world dead, which is why the temple where Albion's Angel finds shelter stands at Bacon's baronial seat, "golden Verulam" (*Eur.*, pl. 10: 5; E, p. 62).[40] It is also why Newton's blast signals an autumnal end, inverting the Resurrection envisioned by Ezekiel:

> Yellow as leaves of Autumn the myriads of Angelic hosts,
> Fell thro' the wintry skies seeking their graves;
> Rattling their hollow bones in howling and lamentation.
>
> (pl. 13: 6–8; E, p. 64)

Newton appears to bring about the death of the material world.

The reverse millennium he provokes is anticipated in *The French Revolution* and in *America*. Louis XVI in *The French Revolution* laments, "The nerves of five thousand years ancestry tremble, shaking the heavens of France; / Throbs of anguish beat on brazen foreheads, they descend and look into their graves. / I see thro' darkness, thro' clouds rolling round me, the spirits of ancient Kings / Shivering over their bleached bones" (4–5: 70–73; E, pp. 285–86). Orc in *America* envisions himself stamping the "stony law . . . to dust: / And [scattering] religion abroad / To the four winds as a torn book, and none shall gather the leaves" (pl. 8: 5–6; E, p. 53).

"But," Orc then adds, their decay shall "make the desarts blossom, and the deeps shrink to their fountains, / And . . . renew the fiery joy, and burst the stony roof" (pl. 8: 7–9). Eyes closed to the infinite open; the millennium is proclaimed to be imminent. If winter comes, spring cannot be far behind.

Nor does *Europe* preclude a revitalized spring. Blake's autumnal energy affirms at least that the Hades bobbin of history—rebirth after death after rebirth after death—will continue. The leaves seeking their graves are comparable to the leaves on which the Fairy proposes to write a book containing those songs that will "shew you all alive / The world, when every particle of dust breathes forth its joy" (*Eur.*, pl. iii: 17–18; E, p. 59). To Blake, as to Wordsworth, nature comprises, for those who can see, characters of the great apocalypse.

And despite the Fairy's mockery of caverned man, the poet-prophet can see. In a world that, to adopt Cowley's metaphor, has rejected "the Desserts of Poetry" (l. 21) for the grapes of Bacon, "prest . . . wisely the Mechanicle way" (l. 75), the poet-prophet not only espies the Fairy but entraps and intoxicates him. The bargain they strike, of which the dictation of *Europe* is the first result, requires that the Fairy be fed "on love-thoughts" and "now and then / A cup of sparkling poetic fancies" to make him "tipsie" (pl. iii: 15–16).

The Fairy, in collaboration with his amanuensis, gives nature a voice, as Orc's embrace gives the daughter (nature spiritualized) a voice. Together, they affirm that the material world lives, is sentient, has consciousness. Consciousness entails pain, particularly from the daughter's birth throes: her fruits maturing to "labour into life" (*Eur.*, pl. 1: 9; E, p. 59). In a fallen state it could hardly be otherwise. *Europe* explores that fallen state, just as *The French Revolution* and *America* explore the moment when that fallen state seems about to pass into eternity. While *The French Revolution* and *America* progress toward an apocalypse, *Europe* retrogresses from an apparent breakthrough capable of triggering the apocalypse (the meeting of poet-prophet and Fairy, with its potential for expanding vision) to immurement in forests of eternal death. Though the birth of Orc into the revolutionary world of *America* alerts Urizen's surrogate, Albion's Angel, to his danger, Orc drives his adversary from the field. The birth of Orc-Christ in *Europe* looses Urizen from his chains (pl. 3: 11; E, p. 60), conferring power on his sons and on the Children of Los and Enitharmon, all of whom are conscripted for the strife of blood. History is reduced to a dialectical process. Winter (Urizen pouring forth his stored snows) has come; and if spring must follow, as day follows night, it seems very far behind.

THREE

Lives of Napoleon: Scott and Hazlitt at Pens' Points

Wrestling with Burke's Ghost

Reading the political signs of the times in 1790, Edmund Burke prophesied Napoleon:

> In the weakness of one kind of authority, and in the fluctuation of all, the officers of an army will remain for some time mutinous and full of faction, until some popular general, who understands the art of conciliating the soldiery, and who possesses the true spirit of command, shall draw the eyes of all men upon himself. Armies will obey him on his personal account. There is no other way of securing military obedience in this state of things. But the moment in which that event shall happen, the person who really commands the army is your master; the master (that is little) of your king, the master of your assembly, the master of your whole republic. (*RRF*, p. 272)

Burke's guide to reading the signs was not universal history, but European, and finally English, history. Alarmed by the consequences he expected the French Revolution to produce, he reminds his reader that "our manners, our civilization, and all the good things which are connected with manners, and with civilization, have depended for ages on two principles ... the spirit of a gentleman and the spirit of religion." In one of the pronouncements of *Reflections* that most outraged English celebrants of 1789, he warns that should the assault on these principles succeed, learning, which he equates with manners and civilization, will, "with its natural protectors and guardians . . . be cast into the mire and trodden down under the hoofs of a swinish multitude" (*RRF*, p. 95). Similarly alarmed by Richard Price's sermon to the assemblage at the Old Jewry, the ostensible occasion for his letter to his young Parisian correspondent, Burke argues that when Price lauded the French Revolution as an affirmation of liberties established for the English by the Revolution of 1688, he picked the wrong precedent:

> That sermon is in a strain which I believe has not been heard in this kingdom, in any of the pulpits that are tolerated or encouraged in it, since the year 1648, when a predecessor of Dr. Price, the Reverend Hugh Peters, made the vault of the king's own chapel at St. James's ring with the honour and privilege of the Saints, who, with the "high praises of God in their mouths, and a *two*-edged sword in their hands, were to execute judgement on the heathen, and the punishments upon the *people*; to bind their *kings* with chains; and their *nobles* with fetters of iron." Few harangues from the pulpit, except in the days of your league in France, or in the days of our solemn league and covenant in England, have ever breathed less of the spirit of moderation than this lecture in the Old Jewry. (*RRF*, pp. 10–11)

Scott, in the preliminary chapters of his *Life of Napoleon Buonaparte* (1827), makes a comparable case, asserting that the signs of upheaval in prerevolutionary Europe were there, for the man able to read them: "accustomed to study the signs of the times, it could not have escaped Frederick [the Great], that sentiments and feelings were afloat, connected with, and fostered by, the spirit of unlimited investigation, which he himself had termed philosophy, such as might soon call upon the sovereigns to arm in a common cause, and ought to prevent them, in the meanwhile, from wasting their strength in mutual struggles, and giving advantage to a common enemy" (1: 19). Translated into historical events (which Scott specifies as the "Flemish disturbances" and the Dutch Revolution of 1787), these sentiments and feelings produce for him other, stronger "signs of the times" (1: 21).

Like Burke, Scott claims that these signs can be read, their predictive potential accurately realized through English history, but that France's revolutionaries and their supporters across the Channel had misread the lesson English history teaches. Although they knew that England beheaded a king and were resolved "that France should not remain behind . . . in the exhibition of a spectacle so interesting and edifying to a people newly regenerated," the French had, Scott laments, chosen to ignore the lesson inherent in the rest of the story: regicide having led to the despotism of Cromwell, England's version of Napoleon, and then to that other British foreshadowing of France's destiny, the Restoration. Had France's revolutionaries followed England's experiment in republicanism to its end, Scott concludes, "they would have obtained a glimpse into futurity, and might have presaged what were to be the consequences of the death of Louis" (1: 177).

Scott is embroidering with historical detail Burke's allegation (almost accusation) that "These gentlemen of the Old Jewry [and their French confreres], in all their reasonings on the Revolution of 1688, have a revolution which happened in England about forty years before, and the late French revolution, so much before their eyes, and in their hearts, that they are constantly confounding all the three together" (RRF, p. 17). His embroidery of Burke highlights the rationale behind Hazlitt's struggle, in his *Life of Napoleon Buonaparte* (1828–1830)— written, he told Frank Medwin, in response to Scott's—to exorcize Burke's ghost.[1] By Hazlitt's measure, Burke's venom toward the "regicide Republic" remained such a strong force in English politics that it caused the rupture of the Peace of Amiens six years after his death. Indeed, Hazlitt represents the voices raised most loudly against the treaty as reviving the case Burke, in *Letters Addressed to a Member of the Present Parliament, on the Proposals for Peace with the Regicide Directory of France,* had made, with all but his dying breath, against a negotiated settlement in 1796 and 1797:

> The British Government and Public at this period [Hazlitt writes of the debate over war and peace in 1803] might be divided into three parties. The first and really preponderating party [the second and third Hazlitt dismisses as the "cat's-paw" and "dupes" of the first] consisted of those

who thought that no peace ought to be concluded with a regicide Republic; and that it was nothing short of national degradation and signing a bond of voluntary infamy to enter into truce or treaty with the traitors and miscreants who had usurped the reins of Government in France, as much as with a den of robbers and murderers whom the laws of God and man made it equally a duty to pursue to extermination or unconditional surrender. This was the high Tory party, the school of Burke and Wyndham, and more particularly including the King's friends. (14: 186)

Numerically the smallest, the school of Burke and Wyndham was, Hazlitt insists, intellectually the most influential of the groups concerned with British policy toward France. To ascribe England's attitude in 1803 largely to a voice silent since 1797 suggests the respect with which, despite his ire at Burke, the apologist for Royalism, Hazlitt continued to view Burke, the political thinker and writer on other subjects, labeled by David Bromwich "the great soul" who orchestrated Hazlitt's inner debates on equality.[2] That Hazlitt has correctly assessed the damage Burke had done through his salvos against the Revolution in France is attested to by Scott, who credits *Reflections* (while conceding its hyperbole) with having "the most striking effect on the public mind of any work in our time," and remarks of Burke himself that "no political prophet ever viewed futurity with a surer ken" (1: 110).

This alertness to Burke's aggressions in England's verbal war over the French Revolution, and to his stature as a political prophet, provoked Hazlitt repeatedly to make him a target of his acid pen, and had done so long before he began his *Life of Napoleon*. As Bromwich puts it, Burke "summoned [Hazlitt's] prose to keener intensities."[3] In "Free Thoughts on Public Affairs: or Advice to a Patriot; in a Letter Addressed to a Member of the Old Opposition" (1806), Hazlitt parodies, in form and phrasing, Burke's *Letters Addressed to a Member of the Present Parliament, on . . . Peace with the Regicide Directory . . .* : "I have promised to say something of the justice of the war in its principle, not as a war of defence but as a war of interference; though I think the less is said on this subject the better; it can only open another Iliad of woes to Europe" (1: 104). In warning that England's interference on the Continent invites "another Iliad of woes," Hazlitt appropriates, and inverts, Burke's own warning—precisely accurate—of the consequences of what seemed to him Westminster's misreading of the geopolitical signs in 1796: "My dear friend, I hold it impossible that these considerations should have escaped the statesmen on both sides of the House of Commons. How a question of peace can be discussed without having them in view I cannot imagine. . . . It opens another Iliad of woes to Europe" (5: 358).[4]

Hazlitt's fury (no softer word will do) at a logic arguing that to make peace threatens "another Iliad of woes," whereas making war promises restoration of an idyllic old order (Burke's happy end to the odyssey of the Bourbons) is underscored by his use of the phrase again, in *The Life of Napoleon*, to stress the extent of the catastrophe engendered by England's violation of the Treaty of Amiens: "Great Britain declared war against France the 18th of May, 1803. Period ever fatal and memorable—commencement of another Iliad of woes not to be forgotten while the world shall last" (14: 197).

The recurrence of this phrase, as almost a refrain in Hazlitt's work, reflects not only his challenge to Burke but also his view of the Napoleonic Wars as an epic in history, ultimately, like the *Iliad*, disasterous for both sides. The life of Napoleon is the story of France in combat with the kings of Europe. The story of France in combat with kings forms an allegory of the birth, growth, struggle, and demise of democracy. France rising to preeminence in Europe, and Napoleon rising to preeminence in France, manifest the strength of the people against a decadent feudal past. France and Napoleon, falling before the armies of Austria, Prussia, Russia, and England, become victims of the Emperor's desire, shared by his subjects, to revive the feudal past.

History, for Hazlitt, teaches ideological lessons; turning history into narration is necessarily a polemical enterprise— which is why he knew as early as 1825, with Scott's *Life of Napoleon* still a mere rumor in the wind, that if it materialized he would answer. In January 1828, having read Scott's *Life* and gotten his own well under way, he wrote a brief essay, "Illustrations of Toryism—from the Writings of Sir Walter Scott," in which he cited excerpts from Scott's sprawling work (nine volumes in its initial printing) to show that even the historian striving for scholarly disinterestedness inevitably lapses into bias; "for if the 'AUTHOR OF WAVERLEY' cannot escape the contagion, who else can hope to be free from it?" (19: 288).

Hazlitt finds the symptoms of that contagion in Scott's rhetoric. He adduces, as the first of four examples, Scott's description of the 1801 concordat between Napoleon and Pius VII, as the Pope surrendering "to a soldier, whose name was five or six years before unheard of in Europe, those high claims to supremacy in spiritual affairs, which his predecessors had maintained for so many ages against the whole potentates of Europe" (in Scott's *Life*, 1: 502). Adding italics to emphasize Scott's discomfort with Napoleon as a general-come-lately to the summit of European power (his *"name was five or six years before unheard of"*), Hazlitt suggests that five or six years of conquest have amply earned the then First Consul his place. He gently mocks Scott for thinking strange the Pope's deference to a man whose sole claim to rank is talent: "It might as well be thought strange, that an author, whose real fame commenced with the publication of 'Waverley' should, in five or six years' time be considered equal to the oldest and greatest writers that ever lived" (19: 289).

Despite his keen eye for the symptoms of contagion in Scott's works, Hazlitt was not himself immune. Scott, had he been so inclined, might have compiled his own list of symptoms, entitled "Illustrations of Radicalism—From the Writings of William Hazlitt." As Hazlitt had begun his scrutiny of Scott with a passage that seemed, in his phrase, "very innocent" (19: 289), Scott might have begun a scrutiny of Hazlitt with a passage that seems equally innocent—the account in Hazlitt's *Life* of his hero's birth. Although Hazlitt read Las Cases's sycophantic memoir of Napoleon, with its tale of the newborn hero laid on a tapestry depicting warriors from the *Iliad*, he describes the infant's first wrap merely as "an old carpet with huge tawdry figures" (13: 4).[5] Having yoked the *Iliad* with England's Burke-inspired vendetta against revolutionary France, and therefore with woe throughout Europe, he dissociates Napoleon from it.

Hazlitt reads—and writes—history symbolically, a practice he shares with Burke himself at one end of the political spectrum and Paine at the other. His narrative of 1789 in *The Life of Napoleon* (chapter 4, "Breaking Out of the French Revolution") thus begins with Marie Antoinette because for him, as very differently for Burke, she epitomizes the times. In 1823, when Scott's *Life* was no more than a faint idea stirring in his mind, Hazlitt had observed, in "Arguing in a Circle" (another of his almost obsessive efforts to exorcise Burke's ghost), that

> the author of the Reflections had seen or dreamt he saw a
> most delightful vision sixteen years before, which had
> thrown his brain into a ferment; and he was determined to
> throw his readers and the world into one, too. It was a
> theme for a copy of verses, or a romance; not for a work in
> which the destinies of mankind were to be weighed. Yet
> she was the Helen that opened another Iliad of woes; and
> the world has paid for the accursed glance at youthful
> beauty in rivers of blood. (19: 272)

Hazlitt is dismantling the most famous (or infamous) of the icons Burke constructed to memorialize the old regime—his portrait of the queen of France, with its associated lament at the decay of a chivalry which would allow such a paragon of beauty to be insulted. The satiric edge Hazlitt applies to Burke's portrait, however, he has honed not with his own rhetorical tools but with tools largely borrowed from Paine's *Rights of Man*:

> As to the tragic paintings by which Mr. Burke has outraged
> his own imagination, and seeks to work upon that of his
> readers, they are very well calculated for theatrical repre-
> sentation, where facts are manufactured for the sake of
> show, and accommodated to produce, through weakness
> and sympathy, a weeping effect. But Mr. Burke should rec-
> ollect that he is writing history, and not *plays*, and that his
> readers will expect truth and not the spouting rant of high-
> toned exclamation. (pp. 286–87)

Hazlitt revives his (and Paine's) campaign against Burke in *The Life of Napoleon*, where he attributes his attack on *Reflections* to one of his minor parliamentary heroes in the cause of peace with France, Charles Fox. Fox, declares Hazlitt, "had not been the dupe of Mr. Burke's romantic and fanciful view of the French revolution, with his high-coloured descriptions of the

Queen of France and the rest of his apparatus for theatrical effect" (15: 274).

In resisting Burke's blandishments, Fox stands as a surrogate for another partisan of the Revolution who refuses to be duped—Hazlitt himself. Wrestling not only with the disembodied author of *Reflections* and *Letters on a Regicide Peace* but with a formidable living adversary in Pitt, Fox is praised for succeeding "as far as it was possible with so disingenuous and artful an opponent, and with the prejudices of his hearers against him" (14: 274). Hazlitt too feels that he wrestles with a disingenuous and artful opponent, and with the prejudices of his readers against him. He claims the power to penetrate his opponent's rhetorical sleight of hand, to expose himself to Burke's "forked shafts" (Bromwich's phrase), yet emerge unscathed.[6] He shakes off, as Fox shook off, "the trammel of words which were attempted to be thrown over him like an enchanter's web" (14: 274).

The trammel of words by which Burke most threatens to enchant unwary minds is contained in his gossamer portrait of the queen of France. As Bromwich suggests, the Marie Antoinette of *Reflections* embodies a fantasy realized in defiance of history. She becomes a tragic heroine in a myth of chivalric ideals tragically crushed by the mob. That the call for a thousand swords to leap from their scabbards in her defense failed to stir the nobles of Europe attests to the eclipse of their glory.[7]

Paine traces the insidious enchantment of this appeal not merely to Burke's trammel of words, nor to the queen, but to the feudal system that (Paine supposes) gave rise to queens. "Titles," he argues, "are like circles drawn up by a magician's wand, to contract the sphere of a man's felicity" (*RM*, p. 320). And he withers Burke for succumbing to their spell: "When we see a man dramatically lamenting in a publication intended to be believed that '*the age of chivalry is gone! that the glory of Europe is extinguished forever! that the unbought grace of life* (if anyone knows what that is), *the cheap defence of nations, the nurse of manly sentiments is gone!*' and all this because the Quixot age of chivalric nonsense is gone, what opinion can we form of his judgement, or what regard can we pay to his facts?" (*RM*, p. 287).

Paine thus anticipates Hazlitt's response in *The Life of Napo-*

leon to a need imposed on him by his polemical design: in debunking the queen called forth by *Reflections*, also to debunk her creator. He posits a Burke even more contemptible than Paine's purveyor of nonsense. Hazlitt's Burke is not simply a man enthralled; he is a turncoat, for he has sold his pen to the enemy:

> Mr. Burke was employed gradually to prepare the public mind for such a change [of Britannia from champion of liberty to ungenerous foe], by sounding the alarm to power and discrediting the popular cause. The loud assertor of American independence appeared first the cautious calumniator, and afterwards, inflamed by opposition and encouraged by patronage, the infuriated denouncer of the French Revolution. He who had talked familiarly of kings as "lovers of low company," now qualified the people as "a swinish multitude." He who had so bespattered the late king that poor Goldsmith was obliged to leave the room, now had occasion to speak of him with proud humility as "his kind and generous benefactor." (13: 50)

This portrayal of Burke as agent provocateur declares, by its very stridency, Hazlitt's respect for Burke's mesmeric skills, which Bromwich characterizes as the capacity to induce readers to identify with the queen, surrendering their interests in favor of hers and accounting themselves generous for having done so.[8] Against that capacity, Hazlitt deploys his own rhetorical stratagems, undermining Burke's metaphors by remolding them to ironic purposes. *Reflections* becomes, in Hazlitt's version, a kind of *Fleurs du mal*, veneering the decay of feudal order: "Mr. Burke strewed the flowers of his rhetoric over the rotten carcase of corruption; by his tropes and figures so dazzled both the ignorant and the learned, that they could not distinguish between anarchy and despotism; gave a romantic and novel air to the whole question; proved that slavery was a very chivalrous and liberal sentiment, that reason and prejudice were at bottom very much akin, that the Queen of France was a very beautiful vision, and that there was nothing as vile and sordid as useful knowledge and practical improvement" (13: 50–51).

In the first chapters of his *Life of Napoleon*, Hazlitt seeks to restore the distinctions that Burke has confounded: to dissipate his romantic and novel airs; to redefine slavery, reason, prejudice as what they are; to portray the queen of France as she was; to reaffirm the virtues of useful knowledge and practical improvement—to redeem, for English readers, the French Revolution. In the beauty Burke perceives as a blessing, ravaged by the mob, Hazlitt perceives a curse, justly recoiling on its embodiment. If Marie Antoinette basks in the admiration of her court (and of enthralled outsiders like Burke), their admiration has made her callous:

> In her the pride of birth was strengthened and rendered impatient of the least restraint by the pride of sex and beauty; and all three together were instrumental in hastening the downfal [sic] of the monarchy. Devoted to the licentious pleasures of a court, she looked both from education and habit on the homely comforts of the people with disgust or indifference; and regarded the distress and poverty which stood in the way of her dissipation with incredulity or loathing. (13: 61)

In narrating the Insurrection of August 10, 1792, when the people in their distress and poverty turned irrevocably against the crown and Marie Antoinette stood more firmly than her vacillating husband, Hazlitt inflates her pride to grotesqueness and reduces Burke's beatific vision to caricature. She is described as accompanying Louis XVI on his last review of the remnant that had rallied to defend the Tuileries, her "Austrian lip and aquiline nose more curled than usual [giving] her an air at once dignified and forbidding" (13: 113).

Hazlitt has lifted this description from Scott, but modified the details to suggest a queen haughty rather than heroic. Scott pronounces Marie Antoinette surrounded by hostile Paris "magnanimous in the highest degree," and in a rare acknowledgment of his source for a historical datum reinforces her claim to heroic stature by calling a witness: " 'Her majestic air,' says Peltier, 'her Austrian lip and aquiline nose, gave her an air of dignity, which can only be conceived by those who beheld her in that trying hour' " (1: 138). Although he restores Burke's (and, in the reverse sense, Hazlitt's) myth to mortal size, Scott makes the tragic queen of *Reflections* the focus for his own lament that chivalry has been extinguished.

He employs the Insurrection of August 10, 1792, as Burke had the march on Versailles of October 6, 1789, when the queen was driven from her bed (where, it was rumored, she was entertaining someone other than the king) by a mob beating down her door. As Burke makes this affront to her dignity the occasion for his interment of European chivalry, Scott makes the storming of the Tuileries on August 10, 1792, the occasion for his: "Alas! instead of the thousand nobles whose swords used to gleam around the monarch at such a crisis, there entered [in the king's defense] but veteran officers of rank, whose strength, though not their spirit, was consumed by years, mixed with boys scarce beyond the age of children, and with men of civil professions, several of whom . . . had now for the first time worn a sword" (1: 138).

Mark A. Weinstein, accentuating the disproportion Scott's Toryism imposes on his narrative, observes that he manages one sentence on the taking of the Bastille, yet lavishes paragraphs on the Royalist resistance in La Vendée.[9] But La Vendée never becomes for Scott more than a skirmish on the outskirts of the Revolution. Weinstein might have made his case against Scott's history still stronger by emphasizing his subordination of the storming of the Bastille to those two other crucial outbursts (which Carlyle a decade later would parallel with the Bastille): by the women of Paris on October 6 and by all Paris on August 10. October 6 and August 10 are, as July 14 is not, crucial to Scott's polemical design because, unlike the attack on the Bastille, the attacks on Versailles and the Tuileries provoked direct confrontations between the people and their king. Embedded in the consequences of October 6, indeed, Scott discovers Louis's ultimate fate foreshadowed: "His wounds salved with this lip-comfort [Bailli's assurance that Louis will gain strength from conquest by his people], the unhappy and degraded Prince was . . . permitted to retire to the Palace of the Tuileries, which, long uninhabited, and almost unfurnished, yawned upon him like the tomb where alone he at length found repose" (1: 84).

Scott too reads—and writes—history symbolically. His simile of the Tuileries as the tomb destined to receive Louis's headless corpse recalls Burke's use of Louis, degraded by his surrender to the mob at Versailles, as a synecdoche for the institution he bespeaks: "The absolute monarchy was at an end. It breathed its last without a groan, without struggle, without convulsion" (*RRF*, p. 166). And in treating the insurrections that destroyed both the absolute monarchy and the constitutional monarchy to follow, Scott presents the queen as a Burkean heroine, unyielding in her elegance. Required to appear alone on the palace balcony to face insults, even a leveled gun, she refuses to flinch, so astonishing the mob with her "noble presence, and graceful firmness of demeanour, [that] there arose, almost in spite of themselves, a general shout of *Vive la Reine!*" (1: 83).

Scott draws this account of the queen's courage from Joseph Weber's *Memoirs of Marie Antoinette* (1805), not (as Carlyle was, with characteristic asperity, to point out in his *French Revolution*), the most reliable of sources. Scott also adds a note, grounded in Madame Campan's scarcely more reliable *Memoirs* (1823), in which he refutes the calumny, as he labels it, that the queen "was on this occasion surprised in the arms of a paramour" (1: 82). He attributes Marie Antoinette's fortitude to her lineage. Called upon to remove her children from the balcony where she stands before the mob, she directs them into the room behind her "with a force of mind worthy of Maria Theresa, her mother" (1: 83).

Worthiness of forebears emerges in Scott's *Life* as a standard of heroism—especially moral heroism—to which the embattled Bourbons repeatedly rise. Invited by Napoleon in 1802 to renounce his claim to the French throne in exchange for a royal sinecure in Italy, the Compte de Provence (Louis XVIII to be) is depicted as refusing with a "firmness of character which corresponded with his illustrious birth and pretentions" (2: 31). Arrested and executed by French authorities in 1804 for conspiring to assassinate Napoleon, the Duc d'Enghien is declared innocent by "all proofs in the case, and especially . . . the sentiments impressed upon him by his grandfather, the Prince of Conde" (2: 44). Menaced by Napoleon's return from Elba in 1815, the Duchesse d'Angoulême (daughter of Louis XVI) is described as showing "the active as well as the passive courage becoming the descendant of a long line of princes" (3: 210).

The Duchess's courage thrusts her among the beleaguered Royalists in Bordeaux, where she is said to comport herself "like one of those heroic women of the age of chivalry, whose looks and words were able in moments of peril to give double edge to men's swords, and double constancy to their hearts" (3: 211). For Scott, however, her firmness attests—as her mother's had on October 6, 1789 and August 10, 1792—to the end of an age in which looks and words from beauty might call forth swords with double edges and hearts of double constancy. His narrative poses her in opposition to Marshal Ney, whose inconstancy of heart (despite having pledged his sword to Louis XVIII, he drew it in the cause of Napoleon) Scott ascribes to an absence of noblesse: "Mareschal Ney was a man of mean birth, who, by the most desperate valour, had risen to the highest ranks in the army. His early education had not endowed him with a delicate sense of honour, or a high feeling of principle, and he had not learned either as he advanced in life" (3: 207).

Napoleon: Orc or Urizen?

A man risen by valor, as well as a host of other talents, and from modest (though not mean) birth to the highest rank, sums up Hazlitt's portrait of Bonaparte. His embrace by the Russian emperor, Alexander, at their meeting on the Niemen in 1807 affirms the common man's equality with kings: "he rose to that height from the level of the people, and thus proved that there was no natural inferiority in one case, no natural superiority in the other" (14: 302). Hazlitt's *Life* is a kind of romance of democracy, with Napoleon as hero. As Bromwich puts it, Hazlitt's campaign against Burke and the seductive power Burke sought, by his trammel of words, to exert over Englishmen entails not disenchantment but "enchantment *on a different principle*": the aura surrounding the queen as heroine is dispelled, then cast anew around a hero identified with the people.[10]

Even as a despot, Napoleon remains in Hazlitt's *Life* the champion of French liberty, who stands against Europe's kings not as Cromwell against Charles I but as William of Orange against James II:

The Allies certainly reckoned on the loose and fluctuating mass of power in France, as the great means of disuniting and subduing it, either by want of concert in the armies or by the collision of different factions. The danger on this side, at least, Buonaparte averted by taking the reins into his own hands, and giving unity and stability to the State; and come what would France thus secured the great principle of the Revolution, the right of changing her existing government for one more congenial to it, like England, which had altered the succession, but retained the forms of her established Constitution. (14: 161)

If the voice is Hazlitt's, the voice behind the voice belongs, again, to Paine, who argues in *The Rights of Man* that although the members of the English Parliament in 1688 had authority to legislate for themselves and their constituents, neither Parliament, nor the Estates-General convened by Louis XIII in 1614, nor the old French parlements could claim the privilege "of binding and controlling posterity to the end of time" (p. 277).

Both Hazlitt and Paine are reacting against Burke's lament in *Reflections* that "the triumph of the victorious party [in France] was over the principles of a British constitution" (p. 166). For them, the triumph of the victorious party was a triumph of the principles—at least those that tended toward republicanism—of the British constitution. And Hazlitt, lauding Napoleon's establishment of unity and stability in France— a political course along which Paine, given the Emperor's methods, would probably not have followed him—anticipates Carlyle, in *On Heroes, Hero-Worship, and the Heroic in History* (1841), lauding Napoleon as an enforcer of order, "the necessary finish of a Sansculottism" (5: 204).

Nothing causes Bromwich, in general Hazlitt's strongest advocate, greater discomfort than the *Life's* unabashed defense of Napoleon the despot. Citing its rationale for Bonaparte's assumption of imperial rank (14: 235)—that monarchy seems "vastly adapted" to human weakness, which, unequal to the demands of choice between right and wrong, welcomes absolutism as a relief from having to choose at all—Bromwich complains that Hazlitt appears to echo Burke, or worse, to have constituted himself a rhetorical kin to one of Marat's Naples bravos, taking up insult in lieu of muff and dirk (or in Bromwich's terror kit, stiletto, truncheon, and other small arms) to achieve the aims of the Revolution.[11]

But Bromwich should not have been surprised to find such weaponry packed into the word-horde of a man who can observe of the execution of the Duc d'Enghien—even while conceding its injustice—that "if it were to do over again, and I were in Buonaparte's place, it should be done twice over" (15: 224). In his survey of Napoleonic historiography, Pieter Geyl makes Enghien's death one of the chief exhibits in the case against Napoleon, who, as dictator and conqueror, sought to impose his will by any means. Having begun his study in Nazi-occupied Holland, Geyl concedes that he is influenced in his assessment of the earlier hungerer after lebensraum by his experience living under the later. He stresses (what has now come to seem obvious) the danger to civilization posed by rulers who practice, unrestrained, the politics of "I want"—though such Hitlerian outrages as the Holocaust had, he acknowledges, no parallel in Napoleonic Europe because, however ruthless, Napoleon's "system remained true, from first to last, to conceptions of civil equality and human rights with which the oppression or extermination of a group, not on account of acts or even of opinions, but of birth and blood, would have been utterly incompatible."[12]

Burke, with the memory of Cromwell and with the Revolution in France before him, reads a comparable lesson in history. By whatever imaginative leap, he finds in the feudal past a system of restraint whose collapse threatens to turn Europe into a Blakean forest of night, where kings and subjects devour and are devoured:

When the old feudal and chivalric spirit of *Fealty*, which by freeing kings from fear, freed both kings and subjects from the precautions of tyranny, shall be extinct in the minds of men, plots and assassinations will be anticipated by preventive murder and preventive confiscation, and that long roll of grim and bloody maxims, which form the political code of all power, not standing on its own honour, and the honour of those who are to obey it. (*RRF*, p. 94)

Both Scott and Hazlitt repeat (one literally, the other ironically) this interpretation of European social history. La Vendée mounted the only sustained resistance to the revolution because, Scott asserts, it was only there that proprietors and peasants "held towards each other their natural and proper relations of patron and client, faithful dependents and generous and affectionate superiors" (1: 27). The subversion that, by alienating proprietors from peasants, has undermined feudalism throughout the rest of France, Scott traces to the crown itself, which, having "concentred within its prerogatives almost the entire liberties of the French nation . . . now, like an overgorged animal of prey, had reason to repeat its fatal voracity, while it lay almost helpless, exposed to the assaults of those it had despoiled" (1: 38).

Hazlitt inverts Scott's analysis of feudal society to arrive at the same conclusion. His peasants look on "their superiors as their natural and declared enemies (whom they had got in their power), not as their natural protectors and benefactors" (13: 75) in a compassionately paternalistic relationship that has been destroyed by a predatory monarchy. In striking anticipation of Dickens, for whom also revolutionary France had become a forest of night, Hazlitt depicts the Sansculottes as men called by the Revolution from "dens and lurking places," and who, "having been regarded as wild beasts, did not at once belie their character" (13: 13). And like Blake, he views the transformation of France into what seems a state of savage nature not as a cause but a consequence of the savagery to which rival kings (devouring and devoured, consumed and consuming) had long since reduced Europe.

Hazlitt memorializes Napoleon in the preface of his *Life* for having stood alone between those kings, uneasily allied against France, and the people, "their natural prey" (13: ix). By his logic, the kings, and their agents plotting to assassinate Napoleon, rather than Napoleon himself, are responsible for Enghien's death: "The guilt did not lie at the door of those who exacted the penalty, but of his own party, who had rendered it necessary by keeping no measures with those whom they chose to regard as outlaws and rebels" (14: 229). Hazlitt, that is, comes to have the same dark vision as Blake: Europe reduced by dynastic rapaciousness to a Darwinian tangled bank where only the fittest (the most ruthless) survive. By making an example of Enghien, Napoleon halts "the annual flight of these bands of harpies, screaming and preparing to pounce on their destined prey" (14: 224).

Hazlitt's conviction, that kingship itself had condemned Europe to an anarchy in which only rules benefiting the rulers applied, enables him complacently to observe of Enghien's execution, "It was contrary to form, I grant; but all forms had been previously and notoriously dispensed with by the opposite party, and an appeal shamelessly made to mere force, fraud, and terror" (14: 225). Actions contrary to form, yet consistent with higher law, sum up for Hazlitt not only the regime of Napoleon but the whole Revolution. He declares the action of the Third Estate in reconstituting itself, "contrary to established rules," as the National Assembly to have been "a stroke of state-necessity to be defended not by forms but by the essence of justice, and by the great ends of humanity" (13: 63). He declares the Mountain and Gironde combining to vote the death of Louis XVI to have been absolved by their establishment of the principle "that if kings presume on being placed above the law to violate their first duties to the people, there is a *justice above the law*, and that rears itself to an equal height with thrones" (13: 129).

That there exists a justice above the law is precisely the proposition Burke and Scott deny. Despite Scott's acknowledgment, in "An Essay on Chivalry" (1818), that the loyalty and freedom of knights often degenerated . . . into tyranny and turmoil,— their generosity and gallantry into hare-brained madness and absurdity," he, like Burke, links the fall of the Bourbons to an end of the generosity and gallantry of France's noblesse in rallying around their king, and their replacement "in France, as in other countries" (1: 25), by a standing army.[13] And he rejects the plea of "state necessity"—entered by Hazlitt to justify not only the Third Estate resolving itself into a National Assembly but Napoleon and his advisers scheming the death of Enghien—as "the Tyrant's plea . . . which has always been at hand to defend, or rather to palliate, the worst crimes of sovereigns" (2: 49).

As Burke puts this idea in *Reflections*, "Kings will be tyrants from policy when subjects are rebels from principle" (p. 94).

Enghien, as a Bourbon in exile, becomes a rebel from principle, "perfectly vindicated," in Scott's view, "by his situation and connexions" (2: 44). Napoleon, occupying the throne of the Bourbons, becomes a tyrant from policy, his execution of Enghien an "indispensable blow" in Hazlitt's view (14: 224), struck to thwart the Royalists' campaign against his rule.

Scott's choice of principle over policy, at least in the rivalry between Napoleon and the Bourbons, is reflected in his treatment of Enghien as a vestige of the generosity and gallantry personified by the knights of romance. Like Henry Morton or Ivanhoe, Enghien is portrayed as being "in the flower of youth, handsome, brave, and high-minded"; unlike Napoleon, making of him an instrument to terrorize future conspirators, he is pronounced merciful and magnanimous: "when [after the Battle of Bortsheim] his army, to whom the French republicans showed no quarter, desired to execute reprisals on the prisoners, he threw himself among them to prevent their violence. 'These men,' he said, 'are Frenchmen—they are unfortunate—I place them under the guardianship of your honour and your humanity' " (2: 42).

None of this dash characterizes Hazlitt's Enghien: "about thirty-four or thirty-five years of age, with light hair, bald forehead, of a good height, and rather corpulent" (14: 225). His description of the Duke is part of the *Life*'s apology for Napoleon as executioner, which argues not Napoleon's innocence but the reasonableness of his error. The police having discovered the plot of Georges, Moreau, and (as yet undetected) Pichegru, "it was naturally concluded that a more important person was concealed somewhere." When the interrogation of Georges's subordinates yields reports of a mysterious visitor, whom Georges had entertained with ceremony, "it was imagined it could be no other than one of the Princes"—an inference pointing, by an inexorable process of elimination, Hazlitt implies, to Enghien: "The description given answered neither to the age of the Count d'Artois nor with the person of the Duke of Berri, whom, besides, Georges's people knew. The Duke d'Angoulême was at Mittau with the Pretender; the Duke of Bourbon in London. There remained only the Duke d'Enghien; and on him the bolt fell" (14: 225).

The use of "naturally," the passive constructions, the appearance of logical rigor suggest—almost as Napoleon's own frequent references to his star suggest—an impersonal force at work in Enghien's death, thus shielding Napoleon and his agents from responsibility. This is not to imply that Hazlitt believed (as some biographers claim that Napoleon believed) in a Hardyan Immanent Will directing history. It is to show how Hazlitt reconciles, for himself perhaps as well as for his readers, his outspoken advocacy of democratic values with his hero's often ruthless exercise of arbitrary power.

Scott denies Enghien's complicity in the plot against Napoleon by a similiar method. The Duke has, Scott alleges, settled in Baden, on the French frontier, "doubtless" to be ready to command "the Royalists in the east of France, or, if occasion should offer, in Paris itself" (2: 42). By "doubtless," Scott concedes that he can cite no proof to support this interpretation. Both he and Hazlitt are building hypothetical constructs, plausible fictions, that are ideologically, not historically, determined.

Hazlitt essentially manufactures an identity of purpose between France's regicides and England's some century and a half earlier, speculating that "the object of both was, as I imagine, to remove the most dangerous enemy of the state" (13: 129). By "as I imagine," he too concedes that, rather than recounting the motives of Cromwell or Robespierre, inferred from their own words or the words of their followers, he projects on them what are, in his revolutionist's mind, morally defensible motives.

Making the guillotine an instrument of moral redress entails a stratagem Carlyle, in his *French Revolution*, was to abominate, sanitizing regicide by depersonalizing it. Paine had, in rebutting Burke, argued that "it was not against Louis XVIth but against the despotic principles of the Government, that the nation revolted" (*RM*, p. 283). Hazlitt adopts Paine's distinction and turns the nation's beheading of Louis into a mainly symbolic act:

> It was not Louis XVI that was properly the subject of debate, but the last remains of arbitrary power, of which he was the representative, the phantom of the past, that rose in irreconcilable antipathy to the prospect of future free-

dom, that no voice could charm, no art could tame; that af-
fecting magnanimity and moderation in public, clung in secret to every vestige of power and prerogative, that shrunk
in fear and loathing from an acknowledgement of the people's rights, and scrupled no treachery, no violence, no
shameless league that promised a chance of finally annul-
ling and disowning them—it was this phantom of kingly
power that was struck at, that tottered and fell headless in
Louis XVI. And with it the opinion, the paralyzing prejudice that that power was sacred, inviolate, and *that* one life
of more consequence than the lives of all other men. (13:
129–30)

Hazlitt's justification of the death of Louis as the demise of
an institution directly challenges Scott's insistence on Louis's
innocence as a man and immunity as a king—suggesting that
Hazlitt wrote much of his *Life* with Scott's open on the table
before him. Hazlitt's rejection of inviolability as a legal principle that protects the lives of kings is an attack on Scott's concurrence in Vergniaud's appeal to the National Assembly that
France, having sworn allegiance to the constitution, "had
thereby sworn the inviolability of the King" (1: 187).

Scott in effect joins Tronchet and Malesherbes as counsel for
Louis's defense. His indictment of the procedures dictating
their client's execution amounts to a legal brief. He denounces
Robespierre's call for the king's death as having "openly disowned the application of legal forms, and written rubrick of
law"; and he associates himself with Vergniaud's case for the
king's inviolability as "truly said," asking rhetorically, "what
right had the convention to protract the King's trial by sending
the case [as Vergniaud had proposed] from before themselves to
the people? If his inviolability had been admitted and sworn to
by the nation, what had the Convention more to do than recognize the inviolability with which the nation had invested the
monarch, and dismiss him from the bar accordingly?" (1: 187)

Scott's response to these questions—that the Convention
had no right to do other than recognize the king's inviolability—is, again, directly refuted by Hazlitt, who claims that "the
effect of this doctrine is to tie the hands of liberty, and to make
men and nations pass under the stroke of despotism, like sheep

under the knife" (13: 129). The simile balances his own hardly flattering, and Scott-like, description of Robespierre's rationale for demanding Louis's head, "that he could not do better, in order to impress on the Revolution that stern, relentless, homicidal character which he wished, than to begin the banquet of blood by the body of an anointed King" (13: 127). Scott himself pronounces Louis passing under the knife a "sacrifice" (1: 191), to which Hazlitt responds that had France not made Louis its sacrifice, the dynasts of Europe would have made France, as well as other men and nations, theirs.

When Scott treats the restoration of the Bourbons in 1814, he turns this whole argument against Napoleon. He labels the emperor's dethronement "a new revolution" (3: 139), justified by the same imperative that justified the dethronement of James II: "The law imposes bounds beyond which the royal authority shall not pass; but it makes no provision for what shall take place, should a monarch, as in the case of James II, transgress the social compact. The constitution averts its eyes from contemplating such an event" (3: 145–46). And Scott claims that Napoleon exiled to Elba "carried away with him all the offences of the French people, like the scape-goat, which the Levitical law directed to be driven into the Wilderness, loaded with the sins of the children of Israel" (3: 176).

Both Scott and Hazlitt seem in their own ways to have anticipated, metaphorically at least, Ronald Paulson's attempt to place revolution under Freudian eyes: his proposal that the incestuous rapes and seductions of which Freud's hysterical patients often accused parents, guardians, or nurses—figures of authority in a children's world—were, as Freud had at first supposed, lived rather than dreamed, and that if an individual might respond to such brutalization by seeking therapy, a people might respond by seeking revenge, sacrificing the king as surrogate father either to absorb his authority or to destroy it.[14] But Hazlitt rejects Scott's assumption that repentant France, in offering Napoleon up as a scapegoat for the sins of the Revolution, is simply the obverse of revolutionary France, offering Louis XVI up as a scapegoat for the sins of the old regime. "The new revolution," he notes, vesting Scott's phrase with a contempt few writers could match, "was represented as being 'a great contract between all the interests of France, in which it was only necessary to sacrifice one interest, namely that of Napoleon'—as if that which was the only stumbling-block to the Allies, was not the only safeguard of France" (15: 206). And Hazlitt dismisses as *"school-boy* cant" Scott's claim (2: 183) that "moderation"—meaning withdrawal to the Rhine—might in 1813 have safeguarded both Napoleon and France. When Napoleon gave the allies "proofs of his moderation," Hazlitt argues (again appropriating his adversary's language for his own ironic purpose), the allies accused him of perfidy: "They called out against his want of plain-dealing and sincerity, with secret treaties and articles of legitimacy in their pockets. They said, 'If you do not come into our proposals, we will accuse you of a desire for eternal war: but the instant you agree to peace, we will break off, insist on terms which we know cannot be granted, and make war on you nevertheless' " (15: 130).

"History with Cloven Hoofs"?

Hazlitt often seems to have written his *Life* convinced that the allies—with Scott, pen in hand, leading the charge—were still making war on Napoleon nearly a decade after his death. To Scott's account of Napoleon in tears at the hostility of the on-lookers along his route through Provence toward Fréjus and the ship waiting to take him to Elba, Hazlitt reacts angrily, pronouncing Scott's version a lie: "It is easy to distinguish the style of the hero from that of the historian: nor is it difficult to understand how a pen, accustomed to describe and to create the highest interest in pure fiction without any foundation at all, should be able to receive and gloss over whatever it pleases as true, with the aid of idle rumor, vulgar prejudice, and servile malice" (15: 215).[15] To Scott's support of the right of Poles in their struggle for independence "to avail themselves of the assistance not only of Napoleon, but of Mahomet or of Satan himself" (2: 126), Hazlitt responds with ironic satisfaction, hailing it as the statement of "a great and admired writer, whose testimony in behalf of liberty is the more to be valued as it is rare" (14: 293).

That Napoleon refused to aid the Poles and thus risk incurring the hostility of Russia and Austria sums up the dilemma his career poses for Hazlitt. It explains both the strenuous effort he mounts to dissociate his *Life* from the *Life* penned by that great and admired writer, and the stubborn resistance of the material itself to his attempts to shape it into an idealized portrait of his hero. The Emperor's "equivocal and calculating policy," as Hazlitt assesses it, leaves him no happier than it leaves Scott, giving him, he admits, "a worse opinion of [his hero] than all he did in Spain" (14: 294). And to Hazlitt, all Napoleon did in Spain sounds " 'the very base string' " (14: 305) of Machiavellianism.

Hazlitt thus finds himself recurrently forced, in the interest of ideological purity, to exaggerate his differences with his Tory rival. He discounts Scott's expressions of admiration for this "most wonderful man" (as Scott had, in the advertisement he wrote for his *Life*, characterized its subject). And when he accuses Scott of creating fiction in depicting Napoleon weeping and trembling "like a woman" (15: 215) en route to Fréjus, the fiction is half Hazlitt's. For Scott's Napoleon weeps, not from fear, as Hazlitt claims, but from the realization that his people have renounced him. In dismissing Scott's version of Napoleon's journey into exile as "history with cloven hoofs" (15: 215), Hazlitt inadvertently reveals the Father of Lies leaving prints on his pages as well.

Hazlitt's conflict with Scott elicits his skill (stressed by Bromwich) at fashioning quotation and allusion into effective weapons for his polemical wars.[16] His complaint that Napoleon's policy toward Spain sounds the base string of Machiavellianism evokes another Machiavellian hero, Prince Hal in *1 Henry IV*, joking with Poins about having "sounded the very base-string of humility" (II, iv, 5–6).[17] Like Hal, Napoleon has degraded himself; like Hal, he also redeems himself.

Borrowing Hal's words to conjoin a mythic hero from England's past with a recent enemy few Englishmen would, except grudgingly, have called heroic, exemplifies what Bromwich (with apologies) labels Hazlitt's rhetorical density.[18] Again and again in his *Life of Napoleon*, Hazlitt uses this device to correct the distortions Scott perpetrates in molding events to a Burkean formula. When Scott remarks on the pause that preceded the colossal Battle of Leipzig in October 1813—that "neither side . . . seemed willing to begin the battle in which the great question was to be decided, whether France should leave other nations to be guided by their own princes, or retain the unnatural supremacy with which she had been invested by the talents of one great soldier" (3: 58)—Hazlitt lifts his words and inverts their message: "Neither side seemed willing to begin a strife which was to decide the great, the only question—Whether the princes of Europe should be put in a situation to dictate laws and a government to France, or fail (as they had so often and so justly hitherto done) incurring the penalty which they madly and wickedly thought this object was worth, not only of disgrace and discomfiture but of their own and their people's subjugation?" (15: 143). When Scott represents the meeting of Alexander and King Joseph of Prussia, after Napoleon's Russian debacle, as a reunion of friends who, though forced to swords' points by European politics, "always retained the same personal attachment for each other," and records Alexander's assurance to the weeping Joseph that "these are the last tears which Napoleon shall cause you to shed" (3: 18), Hazlitt once more lifts his words, but embeds them in mockery:

> The meeting between the two sovereigns was affecting (to them). The King of Prussia wept. "Courage, my brother," said Alexander, "these are the last tears which Napoleon shall cause you to shed." It is to be observed that the tone of these princes was that of persons who were and had always been friends, however necessity or policy might have forced them to dissemble: that as despotic princes they had and could have but one interest at heart, one feeling in common; that whatever appearances they had assumed or engagements they had entered into were merely royal masquerading to conceal or to attain this fixed and favourite purpose; and one of them wept at being assured by the other that this object which had been so long deferred, the restoring the people to their lawful proprietors, had now a chance of being accomplished with the unlooked-for and infuriate acclamations of the people themselves. (15: 120)

Restoring the people to their lawful proprietors echoes, with calculated malice, Scott's description of his feudal ideal, especially as realized in La Vendée. Robert C. Gordon, in an essay tellingly entitled "Scott among the Partisans," discovers the same ideal reflected in Scott's treatment of the Cossacks, who sweep out of the Russian wastes to harass Napoleon's columns, or, "under the eye of their sovereign," repulse Murat's cavalry at Worodonow.[19] Scott self-consciously reverses the conventional prejudice that civilized Europe held the line against barbarian hordes from outside:

> In ancient history, we often read of the inhabitants of the northern regions, impelled by want, and by the desire of exchanging their frozen deserts for the bounties of a more genial climate, breaking forth from their own bleak regions, and, with all the terrors of an avalanche, bursting down upon those of the south. But it was reserved for our generation . . . to see immense hosts of French, Germans, and Italians, leaving their own fruitful, rich, and delightful regions, to carry at once conquest and desolation through the dreary pine forests, swamps, and barren wilderness of Scythia. (2: 324)

Hazlitt resurrects this prejudice in its ancient form and offers it in defense of Napoleon's attack on Russia. To him, "Scythian" is a term of opprobrium. Napoleon invades Scythia because Alexander failed to keep the agreements he signed at Tilsit and Erfurt. Of Napoleon's ambition to dictate peace in the Kremlin, Hazlitt comments portentously, "He knew what resistance civilization could make: did he know equally what resistance barbarism could make?" (15: 18).

What resistance barbarism could make is examined by Count Phillipe-Paul de Segur, whose *Histoire de Napoleon et de la Grande Armée pendant l'année 1812* (1824) both Scott and Hazlitt used in their accounts of the Russian campaign. When Segur describes the advance on Smolensk—which the Russians, warning of things to come, burned rather than leave to the French—he observes, in words a Marxist reflecting on the events of 1812 might have approved, that "It was no longer a war of kings that we were fighting but a class war, a party war, a religious war, a national war—all sorts of wars rolled into one."[20] Hazlitt reiterates Segur's view, tying it, however, to Borodino and asking whether Napoleon's lethargy on the third day of the battle (usually traced to a cold) "did not arise from . . . seeing a still more formidable enemy—the hatred, fear, and despair of a whole people, and the very genius of barbarous desolation standing aghast behind the physical force opposed to him" (15: 52).

This question could have been posed by Scott, who describes Napoleon's invasion of Russia as "hundreds of thousands from the most fair and fertile regions of Europe, moving at the command of a single man, for the purpose of bereaving the wildest country of Europe of its national independence" (2: 324). Scott's description of hundreds of thousands crossing the Niemen at the command of a single man is echoed by Hazlitt, whose chapter "Expedition into Russia" opens with a sentence sprawled over more than two pages (15: 14–16), cataloguing the evils attendant on the allies' struggle against Napoleon and punctuated by a coda, meant to be interpreted ironically: "So much can be done by the wilful infatuation of one country and of one man!" In context Hazlitt allows the identity of that one man (who but Napoleon?) to remain problematical, as he allows his apportionment of blame for the ills of Europe to appear ambiguous. The first in the unfolding series of propositions he appends to his chapter title—"Let a country be so situated as to be able to annoy others at pleasure, but to be itself inaccessible to attack"—describes Russia, as well as England. The elaboration he adds to evoke the ruler of that country—"let it be subject to a head who is governed entirely by his will and passions, and either deprived of or deaf to reason"—applies to Alexander, actually deaf in one ear, and to George III, given to periodic losses of reason. The acts Hazlitt perceives to have sprung from the subjection of that unnamed country to the wayward will and passions of its head—"let it go to war with a neighbouring state wrongfully or for the worst of all possible causes, to overturn the independence of a nation and the liberties of mankind"—suggest Russia's aggression toward Poland and England's toward France.

Hazlitt's summary of the results of aggression—"let it be defeated at first by the spirit and resentment kindled by a wanton and unprovoked attack, and by the sense of shame and irresolution occasioned by the weakness of its pretended motives and the baseness of its real ones"—identifies England as his chief villain. Napoleon ignites this conflict only in the fantasies engendered by the infatuation (madness) that periodically afflicted George III, or by the Francophobia (infatuation in another sense) that had motivated Pitt, dead since 1806, yet seemingly still steering British policy in 1812. A popular general has risen to power in France not, as Burke would have it, because the Revolution was bound to produce him but, Hazlitt insists, because England's hostility toward the Revolution necessitated him: "let it [the England of Burke, Pitt, and Scott] after having exhausted itself in invectives against anarchy and licentiousness, and made a military chieftain necessary to suppress the very evils it had engendered, cry out against despotism and arbitrary sway."

Despite the anger Hazlitt directs at the hypocrisy of Westminster, crying out against the very despotism it has provoked, he must concede hypocrisy's success:

> let it (unsatisfied with calling to its aid all the fury of political prejudice and national hatred) proceed to blacken the character of the only person who can baffle its favourite projects . . . without the least foundation, by the means of a free press, and from the peculiar and almost exclusive pretension of a whole people to morality and virtue; let the deliberate and total disregard of truth and decency produce irritation and ill-blood; let the repeated breach of treaties impose new and harder terms on kings who have no respect to their word and nations who have no will of their own;
> . . . let the uselessness of all that had been done or that is possible to bring about a peace and disarm an unrelenting and unprincipled hostility lead to desperate and impracticable attempts—and the necessary consequence will be that the extreme wrong will assume the appearance of the extreme right; nations groaning under the iron yoke of the victor, and forgetting that they were the aggressors, will only feel that they are the aggrieved party and will endeavour to shake off the humiliation at whatever cost; subjects will make common cause with their rulers to remove the evils which the latter have brought upon them; in the indiscriminate confusion, nations will be attacked that have given no sufficient or immediate provocation, and their resistance will be the signal for a general rising; in the determination not to yield till all is lost the war will be carried

on to a distance and on a scale where success becomes more doubtful at every step, and reverses, from the prodigious extent of means employed, more disastrous and irretrievable; and this without any other change in the object or principles of the war than a perseverance in iniquity and an utter defiance of consequences, the original wrong aggravated a thousand-fold shall turn to the seeming right—impending ruin to assured triumph—and marches to Paris and exterminating manifestos not only gain impunity and forgiveness, but be converted into religious processions, *Te Deums*, and solemn-breathing strains for the deliverance of mankind.

Hazlitt makes Napoleon's Russian adventure an occasion for discrediting the alliance Pitt had forged to defeat first revolutionary and later imperial France. Brunswick's Manifesto of 1792, which denied quarter to all who resisted the Coalition's initial attempt to invade France, is said to proceed from the same rancor as the British press's defamations of Napoleon in 1802 and 1803. The press's vicious campaign against Napoleon is condemned as part of a grand Tory strategy to subvert the Peace of Amiens. And the collapse of the Peace of Amiens released the chaos that rolled east, finally to envelop Moscow, and then turned back until France too—repudiated by many of the same people it had (so Hazlitt asserts) set out to free— "groaned under the iron yoke." He could thus scarcely have repressed his bitterness at Scott's celebration of the crowds that lined the boulevards of Paris on March 31, 1814, to greet the allied kings with shouts of *"Vive l'Empereur Alexandre!—Vive le Roi de Prusse! . . . Vive le Roi—Vive Louis XVIII!—Vivent les Bourbons!"* Nor could he have missed the ironic note in Scott's description (which concludes his chapter on the collapse of French resistance) of the Champs Elysées as "the Hyde Park of Paris . . . converted into a Scythian encampment" (3: 139).

Few modern historians, removed by almost two centuries from the political struggles of both England and France between 1789 and 1815, would, I suspect, accept Hazlitt's radical interpretation of history—positing a conspiracy of kings, organized by Pitt, to deprive innocent France of the right to determine its own future—as wholly credible. For the force of his argument, Hazlitt relies more on rhetoric than on reason. Reacting to the breach of the Treaty of Amiens, Napoleon lays his correspondence with London before the National Assembly to show "that he had done every thing on his part to make good the treaty," which England "had wantonly set aside." Convinced of the justice of its cause, France "warmly" rallies to him and "cheerfully" advances "the means required for issuing victorious out of a struggle in which his enemies could hardly pretend that he was the aggressor, but which was aimed at the existence and independence of the state he governed" (14: 203).

Pretending, or rather (to be neutral) professing, that Napoleon is the aggressor is exactly what his enemies—Scott among them—do. And few modern historians would, I believe, incline more to Scott's Tory interpretation of history—positing a tyrannical France that threatened freedom throughout Europe by challenging the right of people to be ruled by their own monarchs—than to Hazlitt's radical interpretation. Although he purports to rely on reason, embedded in law, rather than on rhetoric (for example, Hazlitt's appeal to higher justice), Scott nimbly, and unselfconsciously, shifts his ground whenever the law fails to serve his brief for counterrevolution. Law may declare Robespierre and his henchmen guilty in the death of Louis, or Napoleon and his counselors guilty in the death of Enghien, but it proves inadequate to judge England for renouncing the Treaty of Amiens: "At common law, if the expression may be used, England was bound instantly to redeem her engagements by ceding these possessions [the Cape of Good Hope, the other Batavian settlements, Malta], and thus fulfilling the articles of the treaty. In equity, she had a good defence; since in policy, for herself and Europe, she was bound to decline the cession at all risks" (2: 19).

The dubiousness of this stance—that some principle of equity, superseding law, may alter agreements between nations— is reflected in Scott's portrayal of Napoleon, remonstrating with the British ambassador, Lord Whitworth, as either madman or Machiavel. In his reconstruction of their meeting at the Tuileries on February 17, 1803, Scott describes Napoleon as talking "incessantly for . . . nearly two hours, not without considerable incoherence" (2: 21). In his reconstruction of their

encounter at a court reception on March 13, 1803, he describes
Napoleon as displaying "a want of decency and dignity" (1: 23).
Scott strains the limits of plausible explanation to reduce Bonaparte to frail mortality, as Hazlitt strains them to raise him beyond mortality into myth. Napoleon's attack on Lord Whitworth, Scott concedes, was more likely prompted by policy
than by pique. When their first meeting has ended, they part,
he allows, "with civility," even if Whitworth is said to have
left convinced "that Buonaparte would never resign his claim
to the possession of Malta" (2: 22).

As Scott well knew, however, Britain, not Bonaparte, subverted the peace by refusing to resign claim to Malta. His real
objection to the then First Consul's treatment of Lord Whitworth is summed up in "his contempt for customary forms"
(2: 21)—the same phrase by which Hazlitt was to justify both
the Revolution and Napoleon. Napoleon's contempt for customary forms, and what Hazlitt, relating the clash with Whitworth, calls "his flaws and starts of temper," identify him as
"still one of the people, and responsible to them for the issue
of affairs" (14: 192). His anger bespeaks neither calculation nor
lack of control, but the natural candor of a common man
charged to protect France, reconstituted by the will of common
men, against the perfidy of kings.

That Whitworth acts out the perfidy of kings, Hazlitt emphasizes by rewriting Scott's version of the meeting at the Tuileries between the Ambassador and First Consul. In Hazlitt's
version, they part "with mutual civility"; but far from entering
a caveat regarding Napoleon's intentions with regard to Malta,
his Whitworth pronounces himself "perfectly satisfied with his
audience" and only later adds a report "tending to inflame the
quarrel and to remove the hope of an adjustment of differences
to a great distance" (14: 19).

Whitworth's perfect satisfaction results, Hazlitt suggests,
from the use he finds he can make of his audience to encourage
resumption of the war. Even some twenty-five years after the
confrontation between the Ambassador and First Consul, Hazlitt continues to reflect the Napoleonic line. In a memoir
(cited by Hazlitt), Savary, Duke of Rovigo, who was one of
Napoleon's close advisers, ascribes the collapse of the Peace of
Amiens to England's scheme to control the Mediterranean by

retaining Malta. Although France, Savary insists, "had faithfully fulfilled her . . . engagements," England refused to accept "The genius of the First Consul, and the . . . prosperity to which he had raised France."[21]

Scott, though he inverts Savary's assignment of responsibility, essentially acknowledges the facts as Savary presents them. "France," Scott argues, "had innovated upon the state of things which existed when the treaty was made, and England might, therefore, in justice claim an equitable right to innovate upon the treaty itself, by refusing to . . . surrender . . . what had been promised in . . . very different circumstances" (2: 19). Between Scott and his republican adversaries, indeed, facts are seldom at issue. For long stretches (his account of the battles of Dresden and Leipzig are good examples), Hazlitt's narrative follows Scott's, often to repeating his words. Their conflict is not over what happened but over how what happened is to be understood. Hazlitt does not deny that French conquests in Italy and Piedmont (what Scott calls France's innovations on the state of things) changed the strategic balance in Europe. But he dismisses as Tory cant the plea that conquests in areas untouched by the treaty entitle England to revise (innovate on) the treaty itself.

So intent is Hazlitt on tracking Scott's case against Napoleon to its Francophobic, and spurious, Tory roots that he follows Scott even in errors of fact. To dramatize the impact of Napoleon's great victory over Austria at Marengo in June 1800, Scott quotes Pitt's lament: "Fold up the map [of Europe] . . . it need not be again opened, for these twenty years" (1: 485). Hazlitt, anticipating his own treatment of Marengo with an enthusiasm Scott could hardly have shared, recasts Pitt's chagrin as celebration: "[Napoleon] had not yet struck though he meditated the blow, which made Mr. Pitt, who had advised and reckoned largely on the continuance of the war, exclaim—'shut up the map of Europe, it will be in vain to open it for twenty years to come!' " (14: 141). Pitt proposed shutting up the map of Europe, however, in reaction not to Marengo but to Austerlitz, observing prophetically that it would be in vain to open it not for twenty years to come but for ten.[22]

Since Marengo and Austerlitz were both milestones in Napoleon's climb to preeminence, we might adopt Hazlitt's view of

Scott as a man who wrote history with a pen accustomed to writing fiction and argue that he has woven Pitt's pronouncement into the narrative, if not accurately, at least appropriately (though this rationale does not support Hazlitt's own claim to have written history with a pen accustomed to history). The conclusion Scott draws from Marengo—"It appeared as if [Napoleon] was the sun of France" (1: 483)—prefigures the sun Napoleon hails, rising from the fog to open the day of Austerlitz, and recalls the title, Sun King, with which he has, all but explicitly, crowned himself.

Scott has in this sense written Marengo with the pen of the historical novelist, who neither manufactures nor misrepresents events but dramatizes them for symbolic effect. He gives Marengo dimensions suggestive, say, of Roncesvalles in *La Chanson de Roland* or (a fiction by a scholar self-consciously imitating *chanson de geste*) the battle at the Pelennor Fields in *The Lord of the Rings*: "The fine plain on which the French were drawn up, seemed lists formed by nature for such an encounter, when the fate of kingdoms was at issue" (1: 481).

Deciding the fate of kingdoms might well mean, as Pitt understood it, changing the map of Europe. Changing the map of Europe might also mean—as revolutionary France understood it—rejecting Europe's past. Fresh from his coup of eighteen Brumaire, and about to launch the campaign that would climax in the victory at Marengo, Napoleon tells Sieyès (in Scott's version), "We are creating a new era . . ." (1: 464). From the perspective of Scott (or Hazlitt) a quarter century or more later, the campaign is easily transformed into a struggle between Europe's past and Europe's future. The Austrian commander, Melas, is eighty; Napoleon is, Scott specifies, "in the very prime of human life" (1: 478). Melas's tactics are mired in old ideas. Napoleon's daring, his contempt for conventional strategy as well as diplomatic custom, enables him to surprise his opponent by crossing Mont Saint Bernard, a route thought too difficult for burdened troops. And when, despite Napoleon's brilliance, Melas seems at Marengo to have won, his strength fails, forcing him to retire, and Napoleon, summoning reserves of energy, simply refuses to acknowledge himself beaten. To Dessaix's concession that "the battle is lost . . . ," Napoleon replies, "By no means . . . the battle is, I trust gained—the disordered troops whom you see are my centre and left, whom I will rally in your rear—Push forward your column" (1: 482).

With God on Our Side?

Scott discerns in the outcome of the Battle of Marengo the conventional fortune of the hero in romance, for whom (again, *The Lord of the Rings* provides an appropriate model) the night is always darkest before dawn. As Aragorn, awaiting a dawn that may never come, warns the Orcs besieging Helm's Deep, "None knows what the new day shall bring him. . . . Get you gone ere it turn to your evil." Scott invokes this same image not to highlight the miracle of Napoleon's triumph at Marengo but to prefigure the turn his boldness takes, to his own, and France's, evil, in the invasion of Russia: "There were . . . few who thought that Russia, in opposition to the whole continent of Europe, would dare confront Napoleon; and still fewer, even of the most sanguine politicians, had any deep-grounded hope that her opposition would be effectual. Out of such a Cimmerian midnight . . . was the dayspring of European liberty destined to arise" (2: 297). The dayspring of European liberty rises out of Cimmerian midnight, for Scott, also in Spain—though in introducing the guerrilla struggle on the Peninsula, he shifts to another, if comparably familiar, image, that of the body politic: "Spain continued that sort of general resistance which seemed to begin after all usual means of regular opposition had failed, as Nature often musters her strength to combat a disease which the medical assistants have pronounced mortal" (2: 302).

This tendency to see history through literary eyes influences Hazlitt as well in his treatment of Marengo, which he adjudges "the most poetical" of Napoleon's battles: "If Ariosto, if a magician had planned a campaign, it could hardly have been fuller of the romantic and incredible. He had given wings to war, hovering like Perseus in the air with borrowed speed. He fell upon his adversary from the clouds, from pathless precipices—and at the very moment of being beaten, recalled victory with a word" (14: 142). Marengo incorporates, for Hazlitt, a romance, an epic, in which Napoleon enacts both Ariosto and Orlando. He becomes the hero of a poem he is himself writing.

Hazlitt foreshadows the Napoleon of Marengo in his portrayal of Napoleon six years younger, leaving the Army of Italy for solitary study in Paris, a youthful Milton in the art of war. This Napoleon is the "man of genius," whose life "from its commencement is a preparation [as Milton's life had been] for the arduous task he has imposed upon himself. His soul is 'like a star and dwells apart,' till it is time for it to disclose itself, and burst through the obscurity that environs it" (13: 183). Hazlitt's simile is lifted, slightly skewed, from Wordsworth's sonnet, "London, 1802," which is addressed to Milton, whose "soul was like a star and dwelt apart" (l. 9): "MILTON! thou shouldst be living at this hour: / England hath need of thee" (ll. 1–2).[23] France, Hazlitt implies, has equal need of Napoleon. But England—beset by apostates like Wordsworth—has, Hazlitt further implies, need of Milton too. To Wordsworth, England in 1802 has need of Milton because its public will is controlled by "selfish men" (l. 6), who have, for their own ease, made peace with the regicide republic. To Hazlitt, England has need of Milton because its poetic voice is monopolized by its Wordsworths (and Scotts), who condemn peace with the French republic.

Hazlitt seems to have written his *Life of Napoleon* in part, and at however late a date, to offer himself as a less glorious Milton on behalf of France's lord protector: on behalf, that is, of both French and English republicanism. His Napoleon emerges as a virtual demigod, almost Christ-like. Indeed, the narrative has features of a saint's life.[24] In his account of the French drive on Mantua in 1796, Hazlitt follows Napoleon into a hospital set up in the Convent of St. Boniface, where, by his presence, he revives two soldiers, "left three days among the dead . . . and . . . recalled to life at the sight of their General" (13: 251). In his account of the return from Elba in 1815, Hazlitt records two proclamations Napoleon issues on the road to Paris, by words resurrecting the nation and its army: "It was the dawn of a brighter day, a raising from the depth of despair, a reprieve from dishonour, a ransom from slavery, a recall from the dead, that seemed little short of miraculous" (15: 228).

Although he declares his hero capable of feats little short of miraculous, Hazlitt consigns to a note Napoleon's own recall from the dead when, as a brevetted second lieutenant, he nearly drowns in the Saone, suffering "all the sensations of dying" (13: 11), only to be washed up on the bank and rescued by his companions. For Hazlitt is, despite appearances, not writing a saint's life. He rejects the providential order saints' lives affirm. When he marshals the signs of Napoleon's imminent fall, he remarks on the paradox of England, which had in 1688 banished a "legitimate sovereign," in 1814 applauding the release of Frenchmen jailed "for attachment to *their legitimate sovereign*": "Thus the centuries stammer and contradict each other" (15: 198).

Scott reads in the restoration of the Bourbons neither stammering, nor contradiction, but divine justice, a step in the mysterious progress of the centuries toward the millennium: "To such unexpected unanimity [as the French display in welcoming the allied kings to Paris] might be applied the words of Clarendon on a similar occasion,—'God had prepared the people, for the thing was done suddenly' " (3: 139). God's preparation of the people to hail the restored Bourbons completes, in Scott's scheme, a cycle begun, if not with 1789, then with the rise to command after 1789 of a regicidal fringe, whose success he rationalizes as "Heaven, in punishment of the sins of France and of the French Revolution, and perhaps to teach mankind a dreadful lesson, [abandoning] the management of the French Revolution" (1: 38).

For heaven to abandon management of the French Revolution is to allow France, and Europe, to recede ever deeper into their forests of night. The irony of life in the forests of night is, as Scott discovers from history, that the law by which one adversary claims the right to conquer another in the struggle for dominance works equally against the conqueror should circumstances change. Examining the aftermath of Napoleon's Russian debacle, Scott justifies Prussia's renunciation of its ties with France in terms Darwinian enough to have satisfied Hardy: "It was by the right of the strongest that France had acquired that influence over Prussia which she exercised so severely, and according to the dictates of common sense and human nature, when the advantage was on Prussia's side, she had a right to regain by strength what she had lost by weakness" (3: 18).

Strength's ruthless suppression of weakness epitomizes, to

Scott, the whole Napoleonic period. Considering France's absorption of the Italian city-states, he borrows Napoleon's own analogy of the alliances between the republic and its dependencies with the embrace of dwarves by a giant: ' "The poor dwarf ... may probably be suffocated in the arms of his friend; but the giant does not mean it, and cannot help it' " (1: 331). Capturing the panic of the French remnants as they struggle—dwarves before the Russian giant—to cross the Berezina River, Scott finds another, especially ironic image of the weak overwhelmed by the strong, not in states but in human beings victimized by rivalries among states: "The weak and helpless either shrunk back from the fray, and sat down to wait their fate at a distance, or mixing in it, were thrust over the bridges, crushed under carriages, cut down perhaps with sabres, or trampled to death under the feet of their countrymen" (2: 384–85).

Hazlitt renders this panic the text for a political homily: a "scene that ... might ... make fiends shudder and kings smile!" (15: 105). Scott too acknowledges, if by inference, that the dissolution of Napoleon's army into the terrified mob packed onto the Berezina bridges might well have made the allied kings smile. But to him, theirs is the justified smile of legitimacy, triumphant over the usurper. Napoleon has, through his war lust, transformed Europe, including France, into a forest of night at which fiends might indeed shudder:

> Wolves, and other savage animals, increased fearfully in the districts ... laid waste by human hands, with ferocity congenial to their own. Thus were the evils, which France had unsparingly inflicted upon Spain, Prussia, Russia, and almost every European nation, terribly retaliated within a few leagues of her own metropolis; and such were the consequences of a system which assuming military force for its sole principle and law, taught the united nations of Europe to repel its aggressions by means yet more formidable in extent than those which had been used in supporting them. (3: 102)

Napoleon himself becomes a casualty of the forest his system has wrought—like Actaeon, the "huntsman ... devoured by his own hounds" (3: 157). Deserted by his marshals in

1814, he is made the victim of a kind of natural selection: "the hurt deer . . . butted from the herd" (3: 264). Earlier victorious at Dresden in 1813, he is "the hunted tiger [springing] upon the victim . . . [in] the circle of hunters, who surround him with protended spears" (3: 43). And still earlier, pursuing his designs against Spain in 1808, he is a snake that has "coiled around the defenceless kingdom the folds of his power" (2: 188).

Constricting defenseless Spain in his folds, he is akin to Robespierre directing the Terror "like a huge Anaconda, [which enveloped] in its coils, and then [crushed] and [swallowed] whatever came in contact with it" (1: 249). Perceived as serpents, both Napoleon and Robespierre resemble Voltaire, who "seduced our imagination," or Rousseau, equally Satanic, "who touched our hearts" (1: 33). For Scott, the French Revolution and its consequences reenact the Fall and the torment the Fall engendered. He traces the roots of the Revolution to the profound change of outlook implanted in human consciousness by "literature—that tree of knowledge of good and evil, which amidst the richest and most wholesome fruits, bears others, fair in show, and sweet to the taste, but having the properties of the most deadly poison" (1: 32).

Such metaphors for the intellectual revolution wrought by the philosophes were suggested to Scott in part by Burke, who habitually likened the development of societies to organic growth, and who warned in *Letters on a Regicide Peace* that "the poison of other states is the food of the new Republic" (5: 238). Scott uses these metaphors primarily, however, to stress that the first victim of this poison is religion, the source of moral order. To clear ground for the *Mai* of French liberty— which, Scott insists, is unrelated to Burke's "British oak" (1: 64)—the philosophes uproot that other "tree [Christianity, the Tree of Life] which bore such goodly fruits" (1: 37).

Uprooting the tree of Christianity entails bending words, and the minds they reach, to perverse ends. And bending words to perverse ends is a skill Scott attributes to Napoleon no less than to Voltaire. When Scott describes the newly named First Consul's efforts to solidify his position after eighteen Brumaire, He portrays Napoleon, Satan-like, assuring Frenchmen of whatever party that " 'All these things will I give you if you will kneel down and worship me' " (1: 464). When he portrays

Napoleon sealing his alliance with Alexander at their meeting on the Niemen, Scott credits him, again Satan-like, with "the sort of eloquence which can make the worse appear the better reason" (2: 141).

This meeting, which resulted in the Treaty of Tilsit, occurred midway through Napoleon's reign, and marked, as Scott and Hazlitt both recognize, the apex of the parabola that graphed his spectacular career. As Scott puts it, in words Hazlitt was to take for his own, "The treaty of Tilsit ended all appearance of opposition to France upon the continent" (2: 146; 14: 304). In fact, Hazlitt would take for his own almost Scott's entire account of the circumstances that led to Tilsit, suppressing only his perception, in Napoleon's control over Alexander, of a diabolic will at work. Hazlitt rejects not alone his rival's indictment of Napoleon as possessing a diabolic will, but the very idea of a will, diabolic, divine, or neutral in the manner of Hardy's Immanent Will, dismissing it as a convenient fiction by which men evade responsibility for what they do—"as if by entailing misery, ignorance, and oppression on a whole nation, it would appear that their degradation and sufferings were in the inevitable order of Providence, and not the effect of our caprice and mismanagement" (13: 26).

Other than chronology, then, the sole structure informing Hazlitt's narrative is the self-consciously rhetorical structure devised by the man of letters, who shapes events to ironic effect. In this endeavor, *Paradise Lost* serves him, as it had Scott. Hazlitt describes Brunswick's Manifesto of 1792 as having " 'like a devilish engine back recoiled' upon its advisors and accomplices," and the liberty their army had set out to destroy as itself the serpent—"a bruised adder," which, menaced by the heel of an Eve shrunk into the hag, Legitimacy, "turned and struck its mortal fangs . . . into those who wished to crush it" (13: 119).[25] He unmasks England as the real Satan, " 'who . . . [during the negotiations of 1814] sat squat like a toad at the ear' of the Allies" (15: 156) to thwart any peace that would consolidate Napoleon's grip on France.[26]

And he mocks what he construes to be England's—and Scott's—hypocrisy in dealing with Napoleon, asking would either have supported a peace that required submission to a hereditary yoke imposed by foreign bayonets "Sixty Years

Since" (15: 156)? Peace imposed by foreign bayonets was seen, at least through Jacobite eyes, as exactly the order Culloden imposed on Scotland "Sixty Years Since." If the yoke to which the Jacobites were forced to bow was not that of the Stuarts, it was—as Edward Waverly painfully learns—nonetheless hereditary and legitimate.

Adjudging *Waverly* (1814) less complex, and less successful, than *Old Mortality*, published two years later, Harry E. Shaw argues that, while both novels call up traumatic episodes from a British past akin to France's present, *Old Mortality* embodies in the Covenanters "the specter" most feared by Scott and his compeers: revolution from beneath, of the sort that had convulsed Britain during the seventeenth century and that on the Continent, after twenty-six years of upheaval, Waterloo had only just brought to a close.[27] The main architect of the counterrevolution, England is, to Scott, the instrument of God, as it is, to Hazlitt, the instrument (metaphorically) of Satan. "Not to that fair land be the praise," Scott proclaims, in one of his frequent excurses on England's glory, "though she supported many burdens and endured great losses; but to Providence who favoured her efforts and strengthened her resolutions; who gave her power to uphold her own good cause, which in truth was that of European independence, and courage to trust in the justice of Heaven, when the odds mustered against her, seemed in earthly calculation, so dreadful as to deprive the wise, of the head to counsel; the brave, of the heart to resist" (2: 296).

Thus Europe is not a Darwinian forest of night; it only seems so to the ungodly. The divine justice that made England a weapon against Napoleon had already foreshadowed his end, when, during his Egyptian campaign he was, like an earlier tyrant to whom Scott compares him, almost drowned in the Red Sea: "The same Deity, who rendered that gulf fatal to Pharoah, had reserved for one, who equally defied and disowned his power, the rocks of an island in the midst of the Atlantic" (1: 419). And the dawn that breaks out of Cimmerian darkness as the Emperor's army disappears under Russian snow imprints on history the lesson of Ecclesiastes: that the maxim, "the race is uniformly to the swift, and the battle to the strong," which Scott pronounces widespread among the courts of Europe, is "as false as it is impious" (2: 398).

Proving the falseness and impiety of that maxim is, Scott declares, the mission to which Providence has called England. He ascribes English involvement on the Peninsula to sympathy for Spain in its fight against overwhelmingly superior French arms. The vow of the defenders of Zaragossa, "war to the knife's blade," confers on Spanish resistance (or so Scott joins the "elegant [and fellow Tory] historian" Robert Southey in proclaiming) " 'a higher and holier character,' " which leads men to regard " 'the issue with faith as well as hope' " (2: 207). Southey and Scott discern in the Peninsular campaign a morality play or, to cite Christopher Dawson, addressing another narrative focused on a battle for Zargossa, a kind of *chanson de geste*, which pits Christian against infidel, good against evil.[28]

Scott's *chanson* produces Napoleon's mighty opposite, Sir Arthur Wellesley, soon to be Duke of Wellington, "one of those gifted individuals upon whom the fate of the world seems to turn" (2: 210). Wellington is God's chosen, in Scott's words, "destined" to an "extraordinary part" by "Providence." He has the breadth of vision of a Blakean prophet: "The past and the future were alike before him; and the deductions derived from a consideration of the whole, were combined in all their bearings, with a truth and simplicity, which seemed the work of intuition, rather than the exercise of judgement." To this "prescience of intellect" he knits physical capacities almost superhuman: "a frame fitted to endure every species of fatigue and privation" (2: 210).

The Wellington of Scott's *Life* flows not from the pen of the historian striving for accuracy but from that of the romancer deploying hyperbole: the "Author of Waverley" as he imagines Henry Morton leading the van against the regulars at Bothwell Bridge—or, ironically, the author of the rival to Scott's *Life* as he depicts Napoleon, say, at Troyes in 1814, like a king of ancient chivalry in the forefront of combat: "Far from shunning the perils of battle, he seemed to court them. He fought at the head of his escort, and was several times obliged to extricate himself from desperate cavalry charges, sword in hand" (15: 190–91). Scott concedes to Napoleon much of the greatness that Hazlitt discovers in him, or that Scott himself discovers in Wellington: "There was," he observes, "considerable resemblance betwixt Napoleon and the English General" (2: 210).

For Scott sees in the Napoleonic conflict, as his frequent use of theatrical metaphors implies, a glimmer of the vast epic drama Hardy was to make of it some three-quarters of a century later, a drama to reach its denouement in the clash of mighty opposites at Waterloo.

Hazlitt, however, finds his mythic analogue to Napoleon neither in Hamlet, nor in any of Shakespeare's other tragic protagonists, but in Sisyphus. Contemplating the collapse of the Emperor's sustained, and heroic, resistance to the allied invasion of France in 1814, he laments that "the ball of victory (which Napoleon had so far endeavoured to roll up its arduous ascent with assiduous pains and dauntless perseverance, and which he had so often suspended on the edge of a precipice by his sole strength and skill) being now left to itself, rolled downhill fast enough with thundering sound to the gates of the capital" (15: 192). This sentence exemplifies Hazlitt's stylistic mastery, by its syntax replicating the very rhythm of Napoleon's Sisyphean labor. It stresses the impossibility of the task he has undertaken against the desertion of him by his own people.

It also stresses the nature of his enemy, especially in Spain and at Waterloo. To Britain's claim (eloquently asserted by Scott) that it has entered the Peninsular struggle to preserve Spanish freedom, Hazlitt opposes an England ("that noted bully and scold") demonstrating itself—at Copenhagen, in the Dardanelles, in Calabria, and in Argentina—impervious to all moral appeal: "I would not believe a word that she said, though she had blown a blast as loud as Orlando's horn at the pass of Roncesvalles, calling on Europe to rise on behalf of Spanish patriotism, liberty, and independence" (14: 332). To Scott's own claim that he perceives in Wellington a genius even greater than Napoleon, Hazlitt opposes a Wellington inert, at Waterloo "in a . . . stupor" (15: 267), saved first by the "courage and obstinacy" (15: 269, 340) he shares with his soldiers, then, when they can hold their ground no longer, by the arrival of Blücher.

In the retrospect of Napoleon's career he offers by way of "Conclusion," Scott too recognizes the Emperor as a type of Sisyphus: "To lay the whole universe prostrate at the foot of France . . . was the gigantic project at which he laboured with such tenacious assiduity. It was the Sisyphean stone which he rolled so high up the hill, that at length he was crushed under its precipitate recoil" (15: 410). But for Scott, that Sisyphean stone is not only England, ultimately represented by Wellington stubbornly making his stand on Mont St.-Jean. It is God's creation, which will not be reordered to suit the designs of a tyrant, however formidable, repeating the sin of Satan. Wellington had to have been more than a stone, triumphant through its own inertia, that Hazlitt accused him of being, because, Scott insists, he was divinely inspired. Abraham Lincoln is said to have told a visitor during the grimmest days of the Civil War that the question is not whether God is on our side but whether we are on God's side. In Scott's reading of history, Wellington and England were on God's side.

FOUR

At the Conflux of Two Eternities: Carlyle's *French Revolution*

The Importance of Knowing Yesterday

Surveying the political and social condition of England in 1829, Carlyle finds it an alarming sign of the times that his countrymen are busy reading the signs of the times: "They deal much in vaticination," he claims in the essay he entitles "Signs of the Times" (*CME* 2: 56). Whatever he knew of Bicheno's pamphlet, *The Signs of the Times,* or of the myriad other oracles, especially on the Revolution in France, it is no coincidence that Carlyle too lifted the title of his essay from Christ's indictment (Matthew 16: 2–4) of the Pharisees and Sadducees asking for a sign: "When it is evening, ye say, It will be fair weather: for the sky is red. And in the morning, it will be foul weather to-day: for the sky is red and lowering. O ye hypocrites, ye can discern the face of the sky; but can ye not discern the signs of the times?" Christ condemns those who profess to need a sign (who lack faith) as "wicked and adulterous." Carlyle, more circumspect, diagnoses the need for signs as "no very good symptom either of nations or individuals" and adds that "our grand business . . . is, not to *see* what lies dimly at a distance, but to *do* what lies clearly at hand" (p. 56).

To do what lies clearly at hand requires discerning, if not portents of a remote future, then "the signs of our own time" (p. 59). That Carlyle accounts discerning the signs of our own time an imperative of moral conduct ("duty") emerges in the quatrain in which he adumbrates his thesis:

Knowest thou *Yesterday*, its aim and reason;
Work'st thou well *To-day*, for worthy things?
Calmly wait the *Morrow's* hidden season.
Need'st not fear what hap soe'er it brings.

(p. 56)

To discern the signs of our own time becomes, then, a mission for the historian, who teaches us to know yesterday, its aim and reason. In "On History," published a year after "Signs of the Times," Carlyle urges the historian to eschew the vaticinator's search for first causes and last effects, "more suitable for omniscience than for human science," and to aim instead "at some picture of things acted" (*CME* 2: 89).

For Carlyle, for all Europe circa 1830, the thing acted that seemed the epitomizing sign of the time was the French Revolution. In *Chartism* (1839), Carlyle characterizes the political agitation in England as "our French Revolution" and prays, with a backward glance at the upheaval in which he had immersed himself for most of the decade, "that we, with our better methods, may be able to transact it by argument alone" (*CME* 4: 149–50). He is reasserting a judgment, expressed ten years earlier in "Signs of the Times," that however ghastly some of the scenes it produced, the overturn of 1789 had an exalted purpose: "The French Revolution ... had something higher in it than cheap bread and a Habeas-corpus act. Here too was an Idea; a Dynamic, not a Mechanic force. It was a struggle, though a blind and at last an insane one, for the infinite, divine nature of Right, of Freedom, of Country" (*CME* 2: 70–71).

By "Mechanic" and "Dynamic," Carlyle implies a distinction, developed in "Characteristics" (1831), between "the mere upper surface that we shape into articulate Thoughts" and the "quiet mysterious depths, [where] dwells what vital force is in us" (*CME* 3: 4–5). The Mechanic is accessible to the scientist, the politician, the manufacturer; the Dynamic is accessible to the artist alone. And the Dynamic generates the currents that make history. "If," Carlyle argues in "Signs of the Times," "we read History with any degree of thoughtfulness, we shall find that the checks and balances of Profit and Loss have never been the grand agents with men; that they have never been roused into deep, thorough, all-pervading efforts by any computable prospect of Profit and Loss, for any visible, finite object; but always for some invisible and infinite one" (*CME* 2: 70–71).

Fortunatus' Hat

Or, as Carlyle asks in *The French Revolution*, "In general, may we not say that the French Revolution lies in the heart and head of every violent-speaking, of every violent-thinking French Man?" (1: 214). Carlyle may have been the first, but he was not the only, historian to see the Revolution as primarily an event of the heart and head translated into the events in the political and social world. Jules Michelet, seeking consolation for the failure of the Revolution to fulfill its promise, tells his readers that "the Revolution lives in ourselves,—in our souls; it has no outward monument." A later historian, Albert Mathiez, declares it to have occurred in the French mind long before it became an observable fact.[1]

Carlyle locates the germ of the Revolution in the psychology of the revolutionaries as a way not only of understanding its dynamics but also, quite self-consciously, of explaining his manipulation of narrative perspective:

> How the Twenty-five Millions of such [violent-speaking, violent-thinking Frenchmen], in their perplexed combination, acting and counteracting may give birth to events; which event successively is the cardinal one; and from what point of vision it may be surveyed: this is a problem. Which problem the best insight, making light from all possible sources, shifting its point of vision whithersoever vision or glimpse of vision can be had, may employ itself in solving; and be well content to solve in some tolerably approximate way. (*Fr. Rev.* 1: 214).

Carlyle is returning to a dilemma that had occupied him during the decade of the *French Revolution*'s gestation and composition. As early as January 1827, he had remarked in one of his notebooks that "an Historian must write (so to speak) in *lines*; but every event is a *superficies*; nay if we search out its *causes*, a solid: hence a primary and almost incurable defect in the art of Narration; which only the very best can so much as approximately remedy."[2] Three years later, recalling in "On History" the experience of Sir Walter Raleigh, who had witnessed a street tumult that he and three others reported differently, he confronts this dilemma again.

Carlyle anticipates the epistemological problem that led Conrad and Ford, among others, to devise a new approach to narrative. And he anticipates their strategy. He shifts his "point of vision," as he calls it, from vantage to vantage, ultimately, by accreting perspectives, to reflect the solidity of events. At its simplest, this strategy entails, for Carlyle, stepping back from the scene to view it "synoptically," as an omniscient narrator—or a conventional historian—might.[3] Conveying the mood of Paris on February 28, 1791 ("The Day of Poignards"), Carlyle moves from Mirabeau struggling to defeat a law against emigration in the National Assembly, to Lafayette restraining tumultuous Saint-Antoine from leveling Vincennes Prison, to royalists with tickets of entry flooding into the Tuileries, and concludes: "Such things can go on simultaneously in one City" (*Fr. Rev.* 2: 131).

This city, the center of the Revolution, is to Carlyle a microcosm, its conflicts a paradigm of what he describes (in *The French Revolution* and elsewhere) as world or universal history. "How much more [can go on] in one Country; in one Planet with its discrepancies," he exclaims as he considers the disruptions to order throughout Paris on "The Day of Poignards." For he takes each day everywhere to consist mainly of disruptions to order, "a mere crackling infinitude of discrepancies, which nevertheless do yield some coherent product, though an infinitesimally small one!" (*Fr. Rev.* 2: 131). In understanding each day as an infinitude of discrepancies which nevertheless yields a coherent product, Carlyle recalls his characterization, in "Signs of the Times," of "the poorest day that passes over us" as "the conflux of two Eternities" (*CME* 2: 59). And he repeats this characterization when he reconstructs the events of May 4, 1789, the coherent product of which was the procession of the Estates-General: "Is not every meanest Day 'the conflux of two Eternities!' " (*Fr. Rev.* 1: 134).

H. M. Leicester highlights "The Procession" in his study of what he terms Carlyle's dialectic of prospective and retrospective telling: his alternation between engaging the reader almost as a character in a dramatic episode and disengaging him to exploit his privileged position, which allows him to comprehend the Revolution whole. The reader is made one with a narrator who—at least from May 10, 1774 (the death of Louis XV) to October 5, 1795 (Napoleon's "whiff of grapeshot")—present, past, and future sees.[4]

The imagination re-creating history works, Carlyle suggests, like Fortunatus' hat, which, as the editor in *Sartor Resartus* (1833) learned from Teufelsdröckh, empowered its owner to wish himself anywhere, and "behold he was there" (p. 260).[5] His way of evoking the scene of May 4, 1789, is to project us into it, inviting us to join him, to "take *our* station also on some coign of vantage; and glance momentarily over this Procession, and this Life-sea; with far other eyes than the rest do,—namely with prophetic" (p. 135). We glance over the procession with prophetic eyes because while in the scene—looking with Baroness de Staël from her window or with the throngs that line the route between Saint Louis Church and Notre Dame—we are not of the scene. We enjoy a kind of dual consciousness that enables us to see in the procession the course events will take, to read it—or, more precisely, to participate in Carlyle's reading of it—as a sign of the times. Of the six hundred marchers our attention is drawn to two: Mirabeau, "as future not yet elected king" (p. 136), and his opposite ("if Mirabeau is the greatest, who of these Six Hundred may be the meanest?" [p. 141]), Robespierre. In them Carlyle asks us to discern the course of the Revolution, from its promise of a new political and social order to its plunge into terror. And as we survey the "Life-sea" around them, we are also asked to recognize faces that would emerge from obscurity only after May 4, 1789: perhaps Louvet standing on tiptoe, or Stanislas Maillard, Captain Hulin of Geneva, Captain Elie of the Queen's Regiment, Jourdan the Headsman, Draper Lecointre, Brewer Santerre of Saint-Antoine, Danton, Camille Desmoulins, Marat.

Carlyle can confirm the whereabouts on this day in this place of none of these men. He suggests their presence interrogatively, speculatively. They should have been there. ("Surely also, in some place not of honour, stands or sprawls up querulous, that he too, though short, may see,—one squalidest bleared mortal, redolent of soot and horse-drugs: Jean Paul

Marat of Neuchatel!" [p. 136].) No historian in the tradition of Ranke or Buckle would, even to capture the atmosphere surrounding an event as crucial as the procession of the Estates-General, have permitted himself such a departure from verifiable fact. Carlyle is manufacturing a historical fiction.

And he adopts a fictive strategy. He establishes a central intelligence (himself as observing eye and oracular voice) to project the reader, dramatically and prophetically, into the action. In an essay focused mainly on *Past and Present*, Elliot L. Gilbert points out that Carlyle restores the past to life by erasing its pastness, by realizing the premise that historical personages vividly portrayed become coeval with us (though about *The French Revolution* it may be more accurate to say that we become coeval with them).[6] As Teufelsdröckh remarks of Fortunatus, "for him there was not Where, but all was Here" (*SR*, p. 261).

To reject "Where" (and When), rendering "all" as "Here" (and Now), is, Gilbert stresses, Carlyle's motive for employing the present tense. *The French Revolution*, Gilbert argues, records the overthrow of a grammatical as well as a political tyranny.[7] To be tyrannized by grammar is, for Carlyle, to distort history. The condemnation detached historians have pronounced on the regicide of Louis XVI is, Carlyle insists, discredited by their detachment: "It is a most lying thing that same Past Tense always" (*Fr. Rev.* 3: 81). The past tense lies because the removal it underscores—of the historian from the event he is recounting—insulates him from the emotional, which is to say from the essential, energy behind it. In the matter of Louis's execution, the historian's removal insulates him from fear: "Not *there* does Fear dwell, nor Uncertainty, nor Anxiety; but it dwells *here*; haunting us, tracking us; running like an accursed ground-discord through all the music-tones of our Existence:—making the Tense a mere Present one!" (p. 81).

The personal pronouns "us" and "our" reinforce the present tense as vehicles for lending immediacy to circumstance. Carlyle is preoccupied with the need to free his imaginative self of the temporal and spatial bounds language imposes on perception, to participate in the past as if it were present.[8] "Be not the slave of Words," Teufelsdröckh urged (four years before *The French Revolution* appeared); "is not the Distant, the Dead, while I love it, and long for it, and mourn for it, Here, in the genuine sense, as truly as the floor I stand on?" (*SR*, p. 55).

The National Assembly making the Constitution, like the historian reflecting on the death of Louis, is for Carlyle a slave of words. While the Assembly labors at what he belittles as its "Theory of Irregular [or Defective] Verbs," the real revolution unfolds in the streets. His denial that the Assembly was a force in the Revolution sets him against prevailing scholarly opinion. For Mathiez and Palmer, to cite historians of different ideologies, parliamentary acts and political maneuvers are crucial. For Carlyle, however, theirs is a mechanical, and external, approach to history. He anticipates R. G. Collingwood, insisting that the historian must get inside his subjects. "History, Collingwood declares, "cannot be scientifically written unless the historian can re-enact in his own mind the experience of the people whose actions he is narrating." Or as he glosses his method in his attempt to comprehend Lord Nelson, "I plunge beneath the surface of my mind and there live a life in which I not merely think about Nelson but am Nelson."[9]

Collingwood essentially argues Carlyle's case for history cast in personal pronouns and present tense. He also describes the creative process as his sort of historian lives it. His idea of how the historian re-creates history is comparable to Keats's idea (gleaned from his attendance at Hazlitt's lectures) of how Shakespeare created drama. The historian writing "scientifically" (in Collingwood's sense) must exercise "negative capability": he must submerge himself in his characters. To write history, especially revolutionary history, entails not analysis but the emplotment of action—in Carlyle's phrase, "an infinite conjugation of the verb *to do*" (*Fr. Rev.* 2: 103).

Toward the Grand Poem of Our Time

To treat history as Palmer or Mathiez treats it, as a "mechanical" rather than a "dynamic" phenomenon, is to rationalize rather than to reenact it, to repeat the delusion of the National Assembly that it makes the Revolution. Carlyle develops this historiographical principle in "Epimenides," an essay on the philosophy of history set midway through *The French Revolution* (Chapter 1 in Book 3 of six, Part 2 of three). By so placing

"Epimenides," Carlyle suggests that it is to be read as an extrapolation from the first half and an orientation to the second half of his narrative. "Epimenides" tells the reader how Carlyle would have *The French Revolution* (the phenomenon as well as the book) understood.

The chapter opens almost as if Carlyle were proposing to answer the question Blake's persona asks the Fairy in *Europe*, "what is the material world, and is it dead?" For Carlyle (inadvertently?) reiterates the Fairy's claim to show us "all alive the world":[10]

> How true, that there is nothing dead in this Universe; that what we call dead is only changed, its forces working in inverse order! "The leaf that lies rotting in moist winds," says one, "has still force; else how could it *rot*?" Our whole Universe is but an infinite Complex of Forces; thousand-fold, from Gravitation up to Thought and Will; man's Freedom environed with Necessity of Nature: in all which nothing at any moment slumbers, but all is forever awake and busy. (2: 102)

That the universe is an infinite complex of forces, and man's freedom within it environed by the necessity of nature, again recalls Carlyle's characterization of each day as the conflux of two eternities. He elaborates this figure in "Signs of the Times," redefining "conflux" as a juncture "of currents that issue from the remotest Past, and flow onwards into the remotest Future" (*CME* 2: 59). Or in the cosmic metaphor he employs in "Epimenides," to image that "All of Things" from which the historian builds his conjugation of the verb *to do*: "From beyond the Star-galaxies, from before the Beginning of Days, it billows and rolls—round *thee*, nay thyself art of it, in this point of Space where thou now standest, in this moment which thy clock measures" (p. 103).

As in Blake, the space that one occupies and the time during which one occupies it merge into space-time. The individual's position in space-time may trigger an epiphany: the perception of "a World in a Grain of Sand . . . Eternity in an hour." It is this grasp of the dynamics of history, accessible not by intellection but by revelation, that Carlyle tried to convey to his friend John Stuart Mill in a letter of September 24, 1833, which displays its author nurturing the germ of his *The French Revolution* and thinking out loud about the monumental task still before him:

> *Understand* me all those sectionary tumults, convention-harangues, guillotine holocausts, Brunswick discomfitures; exhaust me the meaning of it! You *cannot*; for it is a flaming *Reality*; the depths of Eternity look thro' the *chinks* of that so convulsed section of Time;—as thro *all* sections of Time, only to dull eyes not so visibly. To me, it often seems, as if the right *History* (that impossible thing I mean by History) of the French Revolution were the grand Poem of our Time; as if the man who *could* write the *truth* of that, were worth all other writers and singers.[11]

Carlyle is expressing the kind of response to his time that M. H. Abrams finds typical of English radicals. Born largely (as Carlyle was) into nonconformist homes, they tended to retain the chiliasm of their Puritan ancestors. They thus read in the Revolution a sign of a millennial future. And if they grew up to be poets, they gravitated, in Abrams's apt phrase, to forms "suitably grandoise": the epic and Greater Romantic Ode, which posited as their central intelligence a Bard who present, past, and future sees, and incorporated political events into a cosmic frame.

They gravitated, that is, to forms embracing universal history, which, Abrams notes, had for the nineteenth century the status of a literary genre.[12] Carlyle, who claimed that the "right" history of the French Revolution would make the grand poem of his time, suggests as much. He had begun to wonder what this poem might look like even before he embarked on *The French Revolution*. In a notebook entry dated December 1830, he asks himself, "What is a *Whole*? Or how specially, *does* a Poem differ from Prose?" Whereupon he starts, however hesitantly, to grope toward an answer:

> I see some vague outline of what a *Whole* is: also how an individual Delineation may be "informed with the Infinite"; may appear hanging in the universe of Time and Space (partly): in which case is it a Poem and a Whole? Therefore, are the true Heroic Poems of these times to be

written with the *ink of Science*? Were a correct philosophic Biography of a Man (meaning by philosophic *all* that the name can include) the only method of celebrating him? The true History (had we any such, or even generally any dream of such) the true Epic Poem?—I partly begin to surmise so.[13]

Hardy, who (like Dickens) admired Carlyle, saw in *The French Revolution* the poem and the whole for which its author had been striving. Perhaps with his own attempt at an epic on the aftermath of the Revolution already on his mind, he copied into one of his notebooks J. A. Froude's praise of Carlyle's history as "a prose poem with a distinct beginning, a middle, and end."[14]

That *The French Revolution*, any narrative, had to have a distinct beginning, middle, and end posed a vexing problem for Carlyle. To inform an individual delineation with the Infinite is, for the artist, to act heroically—like Carlyle's revolutionary heroes, Mirabeau and Danton, to swallow formulas. But to shape the past according to a pattern that prescribes beginnings, middles, and ends is to be enslaved by formula—to compartmentalize the past and deny Carlyle's idea of how history happens. If each day conjoins currents that flow from the remotest past to the remotest future, the only history that can comprehend the past is one that represents it totally: universal history. In the metaphor Carlyle inherited from Goethe and Coleridge and bequeathed to subsequent generations, history constitutes an organic growth, the germ of which is confined neither to leaf, blossom, nor bole. As he asserts in "Epimenides," "The Beginning holds in it the End, and all that leads thereto; as the acorn does the oak and its fortunes" (p. 103). Or as he asserts early in his narrative, denying the historian's ability to affix blame—specify cause—for the collapse of the old regime: "Friends! it was every scoundrel that had lived, and quack-like pretended to be doing, and been only eating and *mis*doing, in all the provinces of life, as Shoeblack or as Sovereign Lord, each in his degree, from the time of Charlemagne and earlier. All this (for be sure no falsehood perishes, but is as seed sown out to grow) has been storing itself for thousands of years; and now the account-day has come" (*Fr. Rev.* 1: 58).

The apparent contradiction between Carlyle's theory of history and his approach to historical narrative leads Leicester to ask of *The French Revolution* whether it represents "the thing we specifically call French Revolution" or simply molds events, rationalizes them, into some form divorced from the thing.[15] Carlyle recognizes the dilemma. In "On History" he concedes that narration captures "in no case, the real historical Transaction, but only some more or less plausible scheme and theory of the Transaction, or the harmonised result of many such schemes and theories, each varying from the other and all varying from truth" (*CME* 2: 88). In both "On History" and *The French Revolution*, he labels his own attempts to reconstruct historical transactions "approximations."

He is wrestling with a problem that has continued to vex many commentators on the practice of history, if not most practicing historians. With an eye partly on 1789, Claude Lévi-Strauss observes (as Carlyle had before him) that episodes in revolutions or in wars consist of a myriad of individual psychic movements and that no plausible scheme or theory of an episode need be more plausible than any other. Histories are made to seem persuasive not by the historian's ability to verify his findings but by the artfulness with which he builds paradigms, "subsets" of circumstances, that encapsulate periods.[16]

The historian imposes his own conceptualization on events; in Kermode's phrase, he constructs a "fictive concord." Bringing his *French Revolution* to closure, Carlyle acknowledges that he has created a fictive concord. "Homer's Epos," he declares, to open the chapter entitled "Finis," "is like a Bas-Relief sculpture: it does not conclude, but merely ceases. Such, indeed, is the Epos of Universal History" (3: 321). Napoleon's "Whiff of grapeshot" is not the end but merely an ending. By restoring an order that collapsed with the monarchy, it contrasts with the whiff of grapeshot that Broglie, in July 1789, had been unable to deliver. It also allows Carlyle to perceive in the Revolution a progress from a moribund king and court to Mirabeau, France's uncrowned king, but for whose death "the History of France and the World had been different" (2: 138); to Danton, "Mirabeau of the Sansculottes" (2: 205; 3: 47); to the young artillery officer, whose shadowy presence overhangs almost the entire narrative.[17]

That Carlyle views this progress as one of the central themes of *The French Revolution* (phenomenon as well as book) is suggested by his essay "Mirabeau," which appeared the same year as his history: "The French Revolution," he assures readers of the *London and Westminster Review*, "did disclose original men . . . as many as three: Napoleon, Danton, Mirabeau" (*CME* 3: 409). This claim, rhetorically at least, formalizes the judgment he had proferred three years earlier in a letter he wrote to Mill after reading Adolphe Thiers's *History of the French Revolution*: "Three men especially impressed me . . . Mirabeau, Danton, Bonaparte."[18]

Despite Mill's objection that "there were other remarkable men beside those three: Robespierre especially," and despite his own later reassessment, in which he elevated Necker and devalued Mirabeau, Carlyle's original judgment remains manifest in the narrative structure of his *French Revolution*.[19] Napoleon embodies the ultimate triumph of the Carlylean heroic personality in his struggle—waged also by Mirabeau and Danton—to resist chaos. As Carlyle was to put it in *On Heroes and Hero Worship*, "Every Great Man, every genuine man, is . . . a son of Order," for whom involvement in revolutions is "tragical" because "a painful element of anarchy does encumber him at every step,—him to whose whole soul anarchy is hostile, hateful" (5: 203). Carlyle's acknowledgment that in revolutions great men inevitably find themselves encumbered by anarchy leads him, as he tells Mill, to pardon Danton "many things" and to pity him "Heartily at last"—though in the next breath he turns on Thiers for his "Ethics *in petto*," his claim "that the power to have done a thing almost (if not altogether) gave you the right to do it."[20]

Napoleon thus emerges in *The French Revolution* as a moral force: "Let there be Order, were it under the Soldier's Sword: let there be Peace, that the bounty of the Heavens be not split" (3: 316). Carlyle portrays him, taking charge of the defense against insurrection, as the "man at the helm" (p. 319) who pilots the French ship of state over the bar.

This analogy, which recurs in "Vendemaire," the final book of *The French Revolution*, blends Carlyle's view of acted history as a congeries of events and written history as an artifice that conceptualizes events. In *On Heroes and Hero-Worship*,

Carlyle observes of the poet that "He could not sing the Heroic warrior, unless he himself were . . . a heroic warrior too" (p. 79). In *The French Revolution*, Carlyle depicts himself comparably: a pilot steering the ship of his narrative, as Napoleon steers the French ship of state, to safe harbor. "O Reader!—Courage," he exhorts, preparing to bring his treatment of the Revolution to a close, "I see land!" (3: 288).

Death-Birth of a World

The poet—Carlyle writing *The French Revolution*—is heroic because he too is a son of order. The creative process—partly a matter of establishing beginnings and endings—fashions order where none may have existed. As Geoffrey Hartman argues, the narrative artist posits a "cogito" or "Archimedean point," which enables him to make his inner world accessible—to tell a coherent story—to his audience. Or in Edward Said's formulation, to delineate a beginning (Hartman's cogito or Archimedean point) entails delineating an end, satisfying the need to derive unity from experience.[21]

If *The French Revolution* begins with the (literal) death of one king, it ends with the (metaphorical) birth of another. "In rebellious ages," Carlyle asserts, looking back from the platform of his lectures *On Heroes and Hero-Worship* to the lessons he had learned partly through his study of the French Revolution, "when Kingship itself seems dead and abolished, Cromwell, Napoleon step-forth again as Kings. ... The old days are brought back to us; the manner in which Kings were made, and Kingship itself first took rise, is again exhibited in the history of these Two" (p. 204).

Kingship is dead; long live kingship. In one of several myths Carlyle imports to capture the dynamics of the Revolution, he images the upheaval as a Phoenix enacting "the Death-Birth of a World" (*Fr. Rev.* 2: 213). In *On Heroes and Hero-Worship*, he attaches this pattern to his metaphor of hero and artist as pilots steering toward safe harbor, but with the grim caveat that birth after death is not assured:

We will hail the French Revolution, as shipwrecked mariners might the sternest rock, in a world otherwise all of baseless sea and waves. A true Apocalypse, though a terrible one, to this false withered artificial time; testifying once more that Nature is *preter*natural; if not divine, then diabolic; that Semblance is not Reality; that it has to become Reality, or the world will take-fire under it, burn *it* into what it is, namely Nothing. (pp. 201–2)

History describes a series of circles or, potentially, an ascending spiral. The French Revolution encompasses a completed loop, representing in microcosm man's course toward redemption. The Terror, a consequence of mistaking semblance for reality, foreshadows the cosmic, divinely inspired terror that will accompany Judgment:

Slow seemed the Day of Settlement; coming on, all imperceptible, across the bluster and fanfaronade of Courtierisms, Conquering-Heroisms, Most Christian *Grand Monarque*-isms, Well-beloved Pompadourisms; yet behold it was always coming; behold it has come, suddenly, unlooked for by any man! The harvest of long centuries was ripening and whitening so rapidly of late; and now it is grown *white*, and is reaped rapidly, as it were, in one day. (*Fr. Rev.* 3: 202–3)

The harvest, reaped by the guillotine, suggests the harvest Blake evokes in the Prophetic Books, particularly *The Four Zoas*. Both recall the ripe harvest of the earth in Revelation (14: 15–20), cast by the angelic reaper "into the great winepress of the wrath of God" and, in their whiteness, the dry bones that fill the valley of Ezekiel (37: 1–2).

To Carlyle, the French Revolution manifests the hand of Providence guiding history. "Unhappy Sons of Adam," he declaims on the agony of Frenchmen being sucked into the maelstrom, "day after day, and generation after generation, they, calling cheerfully to one another, Well-speed-ye, are at work, *sowing the wind*. And yet, as God lives, *They shall reap the whirlwind*: no other thing, we say, is possible,—since God is a Truth and His World is a Truth" (*Fr. Rev.* 2: 203). Carlyle labels Frenchmen sons of Adam not only because they suffer a terrible, if temporal, judgment but also because, in making their Revolution, they repeat the Fall necessitating Judgment. He traces the pressure behind their growing madness to faith, repeatedly frustrated, "in the possibility, nay certainty and near advent, of a universal Millennium, or reign of Freedom, Equality, Fraternity, wherein man should be brother to man, and sorrow and sin flee away" (*Fr. Rev.* 2: 118).

This faith in the imminent return, by way of revolution, to a Golden Age, Carlyle finds epitomized in the Feast of Pikes. Recounting Paris' feverish efforts to prepare the Champ-de-Mars for the oath of Federation, in a chapter tellingly entitled "As in [not actually "in"] the Age of Gold," he pronounces the mood Edenic: citizens working harmoniously, "distinctions confounded, abolished; as it was in the beginning, when Adam himself delved" (2: 57); all digging and carting (it is July 1790) beneath the evening sun, "a living garden" (p. 59). Carlyle characterizes this immense labor, and the ritual to follow, as a giant theatricality, acted out—like the expedients the government devised to check its slide toward bankruptcy during "the Paper Age" before 1789—in an amphitheater whose borders are the borders of France. The age France ushers in with Federation constitutes only a semblance of gold; it too is paper. Carlyle stresses its flimsiness, the inevitability of its collapse, by setting his scene of idyllic cooperation among classes at evening: "Night is sinking; these Nights, too, into Eternity" (p. 59).

He exploits the vantage he shares with his readers (who, Epimenides-like, look back on the Feast of Pikes and know that it initiated no reign of freedom, equality, fraternity) to give symbolic dimension to his rendering. Because, he assumes, his readers also know the events that compose the rest of his story, to narrate history is less his aim than his instrument for exploring the dynamics of history. In a world, moreover, where nature is preternatural—sentient, possessed of will—to explore the dynamics of history must consist of something beyond establishing chains of cause and effect by methods that purport to be scientific.

Carlyle's method is the poet's method: making the intangible tangible through metaphor. Enriching his narrative through a metaphorical substructure that incorporates truths which transcend mere fact, Carlyle invites the reader to read his book as the prophet-historian reads the "palimpsest" of the past. He demands, as Joyce would almost a century later, that his reader be attuned to the intricate tropic patterns woven into the narrative. When rain soaks the Federates, quenches incense flames, runs the colors on Lafayette's sash, and spoils the ladies' finery, we are induced to recall not only the flimsiness of paper, with which a succession of ministers has sought to shore up the old regime, but the recurrent figures of storm and flood, attaching cosmic import to events. When, two years later, after the slaughter at Nancy and the king's flight to Varennes, the National Assembly reconvenes the Federates on the Champ-de-Mars in an effort to restore the harmony of July 14, 1790, the seats that citizens' labor has built into the slope are again filled, and this time the sun shines, inducing us to recall the spirit of the first Feast of Pikes. "But what avails it?" Carlyle asks rhetorically, and answers, in effect, nothing: "Virtuous Mayor Petion, whom Feuillantism had suspended, was reinstated only last night, by Decree of the Assembly. Men's humour is of the sourest. Men's hats have on them, written in chalk, 'Vive Petion;' and even, 'Petion or Death, Petion ou la Mort.' . . . Sadder Feast of Pikes no man ever saw" (2: 274–75).

The two Feasts of Pikes—one sodden yet celebratory, the other sunlit yet sullen—dramatize man's entrapment in a fallen world, the perversity of which no political contrivance can cure. Why, Carlyle asks, again rhetorically, as he looks a year beyond the first Feast of Pikes to the bloody confrontation between Feuillantism and Sansculottism also to take place on the Champ-de-Mars, should an oath, "sworn with such heart-effusion, emphasis, and expenditure of joyance," be so soon irremediably broken? "Chiefly," he answers, because "Sin had come into the world and Misery by Sin!" (2: 68).

Paradise Lost

To Carlyle, the French Revolution reaffirms—Miltonic style—the Christian theory of history (which may be what he meant when he confided to his notebook that equation between "true" history and "true" epic). Contemplating the acrimony in store for the principals in the marriage between royalty and the nation, celebrated at the first Feast of Pikes as promising (like the marriage of Christ and his church) new heaven, new earth, he laments: "Fond pair! the more triumphant ye feel, and victorious over terrestrial evil, which seems all abolished, the wider-eyed will your disappointment be to find terrestrial evil still extant. 'And why extant?' will each of you cry: 'Because my false mate has played the traitor: evil was abolished; I, for one, meant faithfully, and did, or would have done' " (2: 68). Carlyle's drama of the bickering between marriage partners so uneasily joined echoes the bickering of another fond pair as they, having transgressed God's law, await punishment.

Carlyle extends this marriage figure when he asks (once more rhetorically), "Shall we say then, the French Nation has led Royalty, or wooed and teased poor Royalty to lead *her*, to the hymeneal Fatherland's Altar, in such over-sweet manner; and has, most thoughtlessly, to celebrate the nuptials with due shine and demonstration,—burnt her bed?" (2: 69). This question ends Book 1 of Part 2 ("The Constitution"). A scene focused on another burning bed, the death bed of Louis XV, ends Book 1 of Part 1 ("The Bastille"). Carlyle finds (or fabricates?) order within the apparent chaos of events. He attests to his discovery by paralleling episodes, thereby giving structure to the narrative and claiming form for the Revolution. The bed into which royalty is seduced by the nation suggests the "Bed of Justice" (1: 91) into which, even in the face of its own bankruptcy and Orleans's challenge, royalty had failed to compel the Parlement of Paris in 1787. It also suggests—an image that specifically evokes Louis XV's death throes—the bed toward which royalty is said to march when forced from Versailles to Paris by the mob present at the Insurrection of Women.

Carlyle represents the progress of the king's carriage and the throngs that escort it back to Paris as one in a series of processions that punctuate French, indeed Western, history: "Processional marches not a few our world has seen; Roman triumphs and ovations, Cabiric cymbal-beatings, Royal progresses, Irish funerals; but this of the French Monarchy marching to its bed remained to be seen" (1: 287).[22] Describing the bedward march of the French monarchy as remaining to be seen identifies the King's return to Paris as both a crucial stage in his march toward dethronement and a foreshadowing of the steps by which he mounts to the guillotine, those "Two other Paris Processions [he has] to make: one ludicrous-ignominious like this:

the other not ludicrous nor ignominious, but serious, nay sublime" (1: 298).

The allusion to Louis XVI's last procession, in a tumbrel rumbling over the stones to the Place de la Révolution, recalls Carlyle's brief reversion to Louis's distant forebears, in whom are summed up the fate of the Bourbons, of all dynasties, the flow of those currents from the remotest past to the remotest future: "Sovereigns die and Sovereignties: how all dies, and is for a Time only; is a 'Time-phantasm, yet reckons itself real!' The Merovingian Kings, slowly wending on their bullock-carts through the streets of Paris . . . have all wended slowly on,— into Eternity" (1: 7).

The Merovingian kings wending through Paris adumbrate not only the processions which will subvert the power they had accrued, and that had been wielded absolutely by Louis XIV (L'état, c'est moi), but also the garish display of wealth and privilege through which the French nobility, exhibiting its decadence, invites its own destruction. Appropriating an episode from Mercier's Tableau de Paris, Carlyle depicts a parade of spring finery, "manifold, bright-tinted, glittering with gold," in the Bois de Boulogne, its participants indifferent to the struggle of the government (it is 1783) to fend off insolvency, as well as to the struggle of the masses to fend off starvation. He labels this parade too a "procession: steady, of firm assurance, as if it rolled on adamant and the foundations of the world; not on mere heraldic parchment,—under which smoulders a lake of fire" (1: 48).

Heraldic parchment specifies metaphorically the "paper" substance of the age, defined also by the title of the book ("The Paper Age") in which the scene occurs. The lake of fire beneath the parchment again suggests Milton, casting those strollers in the Bois de Boulogne as Satanic cohorts about to be hurled into the abyss. The setting in which they stroll, "like long-drawn living flower-borders, tulips, dahlias, lillies of the valley" (1: 48), is a garden—what Carlyle, never embarrassed by a cliché if he found it rhetorically useful, might have called a fool's paradise. It anticipates another fool's paradise, the Eden of the Champ-de-Mars in July 1790, and a paradise irretrievably lost: the Tuileries after the insurrection of August 10, 1792, its grounds strewn with dead Swiss, "like no Garden of the Earth" (2: 303).

"Ye and your fathers have sown the wind," Carlyle in his prophetic voice addresses France's nobility, as he would later address their countrymen enmeshed in the Terror, "ye shall reap the whirlwind" (1: 48). Their procession, rolling and dancing through the Bois de Boulogne, yields to "Carmagnole complete," Sansculottism's dance of death.

Carlylean Palimpsest

Charles Frederick Harrold might have cited this scene as evidence when he long ago described Carlyle's method in *The French Revolution* as a selection and transformation of passages lifted from his sources and fit into a "vast word-picture."[23] For "transformation," Harrold might also have substituted "translation" or, in cases where texts were available in English, "paraphrase," which Carlyle then disguises with the rich metaphorical embroidery that makes his sources his own. Thus he draws his account of Louis XV's death from Madame Campan's *Memoirs of Marie Antoinette*. But to suggest that "Newton and Newton's Dog Diamond" perceive different universes, and that "the Reader here, in this sickroom of Louis, [must] endeavour to look with the mind too" (1: 5), he alters Campan's tone. Her version of the scene exudes delicacy and deference:

> The dauplin had settled that he would leave [Versailles for Choisy] with the royal family, the moment the King should breathe his last sigh. But upon such an occasion, decency forbade that positive orders for departure should be passed from mouth to mouth. The keepers of the stable, therefore, agreed with the people who were in the King's room that the latter should place a lighted taper near a window, and that at the instant of the King's decease, one of them should extinguish it.
>
> The taper was extinguished. On this signal the bodyguards, pages, and equeries [*sic*] mounted on horseback, and all was ready for setting off.[24]

In Carlyle's version everyone is gripped by an indecorous haste to be gone: "In their remote apartments, Dauphin and Dauphiness stand road-ready; all grooms and equerries booted and spurred: waiting for some signal to escape the house of pestilence" (1: 24). And the signal Campan reports them awaiting, Carlyle in a note reasons away with mockery:

> One grudges to interfere with the beautiful theatrical "candle," which Madame Campan . . . has lit on this occasion, and blown out at the moment of death. What candles might be lit or blown out, in so large an Establishment as that of Versailles, no man at such distance would like to affirm: at the same time, as it was two o'clock in a May Afternoon, and these royal Stables must have been five or six hundred yards from the royal sickroom, the "candle" does threaten to go out in spite of us. (1: 24–25)

The signal Carlyle reports is a " 'sound terrible and absolutely like thunder' . . . the rush of the whole Court, rushing as in wager, to salute the new Sovereigns" (p. 25). Although Campan too records this thunder—Carlyle's phrase is Campan's translated—she mutes its implication of sycophancy, which he accentuates by giving it speech: "Hail to your Majesties!" (p. 25). She also provides Carlyle with the portrait of fledgling king and queen in tears, on their knees, praying for God's guidance and protection because they feel themselves too young to reign. But what for Campan is an expression of piety and grief is for Carlyle a portent, the irony of which he underscores by his own addendum, "Too young indeed" (p. 25).

Superimposing his voice on Campan's—as well as on the myriad other writers he calls up to capture France in revolution—is the device Carlyle uses to explore the gap between "the eye of History," to whom "many things, in that sickroom of Louis, are now visible," and the eyes of "The Courtiers there present, [to whom those things] were invisible" (1: 5). The death of Louis XV—dramatized, to my knowledge, in no account of the Revolution except Carlyle's—adumbrates his rhetorical strategy. It defines a frame of reference, orienting the reader to the epistemological, the interpretive, issues that influence the way the Revolution is rendered.

In an essay entitled "Carlyle and the Fictions of Belief: *Sartor Resartus* to *Past and Present*," Janet Ray Edwards maintains that the mask Carlyle invents for himself, of editor (or historian?) immersed in his sources, reinforces the vehicle of present tense narration by which he transports the reader into events.[25]

It seems to me, however, that by donning this mask, Carlyle enables the reader to occupy both realms at once—to stand among the courtiers at the dying king's bed while looking also through history's eye, seeing those things that remained invisible to the courtiers themselves.

Carlyle had been arguing as early as 1830 that such a multiplicity of perspectives is essential to achieving even a plausible approximation of the past. He labels the alternative versions of the street tumult recounted by Raleigh and three other witnesses "a true lesson for us" (*CME* 2: 87). He reiterates the truth of the lesson in the "Finis" to *The French Revolution* through his otherwise inexplicable device of giving the last apocalyptic word to the archquack, Cagliostro. Cagliostro's vision of "imposture . . . burnt up" (3: 322) is an ultimate irony, for he is himself an imposture. When Carlyle returns to ask, in his own voice, "This Prophecy . . . has it not been fulfilled, it is not fulfilling?" (p. 323), he implies that the prophet too, false as he is, will follow king, queen, Iscariot Egalité, De Launay, and the Bastille, "whole kindreds and peoples" (p. 323), into the fire.

The Revolution shows Cagliostro his own destiny. Right for no real reasons, he manifests God's power to use even the unlikeliest instrument to direct history toward completion of its appointed rounds. Cagliostro's Jeremiad at the end balances Carlyle's indictment of Louis XV's court as an "enchanted Dubarrydom" (1: 3) at the beginning. Dominated by an incarnation of the Great Whore, French kingship has been reduced to a conjurer's trick.

Carlyle emphasizes Louis XV's desire to live in a world of conjurer's tricks, to adopt "the resource of the Ostrich" (1: 19), through another incident, drawn from Campan, and expressing the king's cultivation of the most illusory of all human illusions, that he can evade death: "Hunting one day in the forest of Sanard, in a year in which bread was extremely dear," so Campan's version runs, "he met a man on horseback, carrying a coffin. 'Whither are you carrying that coffin?' 'To the village of ***,' answered the peasant. 'Is it for a man or a woman?'— 'For a man.'—'What did he die of?'—'Hunger,' bluntly replied the villager. The King spurred his horse, and asked no more questions."[26] Though Campan acknowledges that Louis's abrupt departure bespeaks his pathological fear of death, she mutes its impact and deflects attention to his royal dignity and excellent health, relegating the encounter to a kind of appendix ("Anecdotes of the Reign of Louis XV").

Carlyle integrates the encounter into his narrative and thus amplifies its impact. Campan's diagnosis, that Louis "was extremely apprehensive of death," he enlarges to: "Louis XV had always the kingliest abhorrence of Death" (1: 19); her claim that because he feared death, he "liked to talk about [it]," Carlyle inverts: "He would not suffer Death to be spoken of; avoided the sight of churchyards, funeral monuments, and whatsoever could bring it to mind" (p. 19). Carlyle also intensifies the confrontation between boundless privilege and abject poverty. While Campan's harbinger of death is a man riding a horse, Carlyle's is "a ragged Peasant" (p. 19), and the horse has disappeared; while her victim is an identityless corpse, whose struggle for life has gone on elsewhere, his is "a poor brother slave, whom Majesty had sometimes noticed slaving in those quarters" (p. 19).

Carlyle remolds Campan's scene into an allegorical vignette, incorporating what he takes to be some of the chief causes of the Revolution. The hunt evolves, indeed, into a metaphor for the irresponsibility that characterized Louis XV and that he passed on fatally to his son. Of Louis XV, occupied with his enchanted Dubarrydom, Carlyle asks, "Who is it that the King . . . now guides?" and answers: "His own huntsmen and prickers: when there is to be no hunt, it is well said, 'Le Roi ne fera rien (Today his Majesty will do nothing.)' " (1: 11–12).

Doing nothing sums up Louis XVI. Challenged by the turmoil attendant on the destruction of the Bastille, Louis can, Carlyle asserts, resolve only to avoid civil war: "For the rest, he still hunts, having ceased lockmaking; he still dozes and digests; is clay in the hands of the potter" (1: 245). As the Insurrection of Women marches on Versailles, he is in fact hunting in the Woods of Meudon; and hunting in the Woods of Meudon fulfills the wish embedded in the dream Carlyle—again using fiction to illumine history—speculates that "Majesty, kept in happy ignorance, perhaps dreams" (1: 200) while the Bastille is being stormed.

The hunt in its recurrent figurations measures the lapse into impotence not merely of Louis XVI but of the French monarchy. As Carlyle specifies, telling us how to read the death of Louis XV, "not the French King only, but the French Kingship; this too, after long rough tear and wear, is breaking down" (1: 7). The extent of its breakdown, even before 1789, is acted out by Louis XVI when, having announced a royal hunt for the morning of November 19, 1787, he starts his day by demanding that the Parlement of Paris register his edicts to tax the aristocracy. And Carlyle underscores the meaning of his failure: "What a change, since Louis XIV entered here, in boots; and, whip in hand, ordered his registering to be done,—with an Olympian look, which none durst gainsay; and did, without stratagem, in such unceremonious fashion, hunt as well as register!" (1: 90).

It is but a short step from Louis XVI's entrance into the Parlement of Paris in futile search "of two-legged unfeathered game" (p. 90) to his confinement in the Tuileries after the Insurrection of Women. Now he "can get no hunting. Alas, no hunting henceforth; only a fatal being-hunted!" (2: 4). Being hunted reduces him to one more among French nobles, whom Sansculottism had already reduced to two-legged unfeathered game and taken to hunting "as if they were wild wolves" (2: 13).

This decline from hunter to hunted climaxes in the overturn of August 10, 1792. That Carlyle detects providential justice at work in the decline and its consequences is attested to by the symmetry he finds in—and builds into his treatment of—events. As the near sack of Versailles on October 5, 1789, was preceded and in part provoked by the dinner, with its pledges of loyalty to the king and especially the queen, that the Gardes-du-Corps had hosted for the Regiment-de-Flandre on October 1, so the actual sack of the Tuileries on August 10, 1792, was preceded and in part provoked by the dinner, with its pledges of loyalty to the nation, that Saint-Antoine hosted for the Marseillais on July 30. The near sack of Versailles, which forced the king's removal to Paris, ends Part I, "The Bastille." The actual sack of the Tuileries, which forced the king's removal from the throne, ends Part II, "The Constitution."

Carlyle derives his account of the king's downfall largely from Syndic P. L. Roederer's *Chronique de Cinquante Jours.*

But he alters Roederer's point of view. Roederer's chronicle of the last fifty days of Louis XVI's reign is mostly a narrative of Roederer himself standing heroically by his king. When Louis decides, at his syndic's urging, to quit the Tuileries for the Assembly, the queen's sister, Elizabeth, asks whether Roederer will answer for the king's life ("Monsieur Roederer, vous répondez de la vie du roi?"), to which he replies, "Yes, madame, with my own" ("sur la mienne"). He is rewarded (or so he supposes) with a look of confidence from his monarch ("Le roi me jeta un regard du confiance"). When they approach the Salle de Manège escorted by their faithful syndic (and by the reader, whom Roederer's first-person narration makes one of the procession), the royal family must brave a "Fremescent multitude" (Carlyle's phrase [2: 297]), which includes a man wielding a long pole; whereupon Roederer, by his own account, takes charge: he quells the mob with a call for silence in the name of the law, wrenches the pole from its owner, and throws it into the garden.[27]

Much of Roederer's effort to fix himself at the center of Louis XVI's losing struggle Carlyle either omits or alters. The wielder of the long pole is, in Carlyle's version, "stilled by oratory" (2: 298)—whose remains unspecified. The king's progress across the garden to the Salle de Manège unfolds through the eyes not of Roederer as he clears the way but of the Swiss and Louis's loyal remnants ("gallant gentlemen in black" [p. 297]), who watch from the windows of the Tuileries, where they wait to be massacred.

Carlyle changes the point of view of Roederer's *Chronique*, though he retains much of its detail, in the interest of a precision both historical and symbolic. He repeatedly reminds us that the memoirs he quarries to reconstruct the experience of the Revolution are self-serving, unreliable. And the need to improve on history was not, he emphasizes, felt only at court. Camille Desmoulins at the hour of his trial was thirty-four rather than, as he would have it, thirty-three, the age "of the *bon Sansculotte Jésus*" (3: 257). The *Vengeur*, sunk by a British man-of-war on June 1, 1794, did not, despite the grandiose claim in Barère's eulogy, go down with tricolor flying and all hands on the upper deck shouting *Vive la République* (though that piece of what Carlyle calls inspired *blague* proved inspiring enough to fool him, and had to be recanted in a later edition of *The French Revolution*).

Carlyle discerns in these fabrications a species of myth-making, and he accords them grudging respect.[28] In Roederer, however, he sees only a political trimmer caught between the proverbial rock and hard place: "Will the kind Heavens open no middle-course of refuge for a poor Syndic who halts between two?" (2: 296). Halting between two describes the poor syndic's stance before the troops manning cannons in defense of the Tuileries and asking whether they must fire on the people (our brothers, "nos frères," in Roederer's version); to which he answers that they need only guard the doors where they are posted and fire if fired upon ("Vous n'êtes là que pour garder cette porte, empêcher qu'on n'y entre; vous ne tirerez qu'autant qu'on tirerait sur vous: si l'on tirait sur vous, alors ce ne seraient pas vos frères").[29]

Carlyle, while he hews closely to Roederer's narrative line, expunges his reassuring words, substituting for them his own irony: "Syndic Roederer has a hard game to play. He speaks to the Cannoneers with eloquence, with fervour; such fervour as a man can, who has to blow hot and cold in one breath" (2: 297). Representing Roederer by a paraphrase that reduces him to a political caricature is one of Carlyle's stratagems for moving him from the center of events to the periphery, for refocusing attention on Louis XVI. For, as Carlyle implies through the death scene of Louis XV, the king embodies the forces—both virtuous and vicious—that make the Revolution inevitable. He is one of a group of figures who, while part of the history Carlyle narrates, are also worked into an allegorical substructure he invents. Shifting the perspective on Louis XVI's abandonment of the Tuileries from Roederer to the royalists who are left behind enables him to render the event a symbol of the collapse of Royalism itself: "Royalty has vanished forever from your eyes.—And ye? left standing there, amid yawning abysses, and earthquake of Insurrection . . . if ye perish, it must be . . . as martyrs who are now without a cause" (2: 298).

The condemnation of Louis XVI's remaining supporters to causeless martyrdom represents the end of a process as inexorable in its advance (tracked by the sequence of processions) as the movement of a Newtonian object through a vacuum. When (in the "Procession of the Black Breeches") the mob, gathered on June 20, 1792, to plant a tree of liberty, floods the Tuileries, its invasion of royalty's private quarters is treated as a precursor of the events of August 10. The encounter between Louis and his people is allegorized as "Incongruity fronting Incongruity, and . . . recognizing themselves incongruous" (2: 261–62).

To Carlyle's historical (hence symbolic) understanding, this confrontation is crucial. It heralds the breakdown not merely of one dynasty, that of the Bourbons, but of the whole European dynastic system: "Serene Highnesses, who sit there protocolling and manifesting, and consoling mankind! how were it if, for once in a thousand years, your parchments, formularies and reasons of state were blown to the four winds; and Reality Sans-indispensables stared you, even you, in the face; and Mankind said for itself what the thing was that would console it?" (2: 279).

Mankind staring Serene Highness in the face and trying, however inarticulately, to say what would console it defines the meaning Carlyle perceives in the moment on June 20 when Louis XVI, the mob pounding at his door, opens it to ask, "What do you want?" Reproducing the mob's reply, "Veto! Patriot Ministers! Remove Veto!" and Louis's (supposed) answer, "this is not the time to do, nor this the way to ask him to do" (2: 262), Carlyle alters the scene he found in Roederer to highlight the drama of Incongruity fronting Incongruity. In the *Chronique de Cinquante Jours*, it is not Louis but Max Isnard who, crossing from the Salle de Manège, quiets the people, and admonishes them that this is neither the time nor the way to petition the king.[30]

Carlyle pronounces Louis's steadiness before the mob "valiant" and enjoins the reader to "honour what virtue is in a man. Louis does not want courage; he has even the higher kind called moral-courage, though only the passive-half of that" (p. 262). Carlyle further alters Roederer's version to stress Louis's moral courage. In the *Chronique*, the exchange between the member of his guard (identified by Roederer as M. de la Chesnaye) and Louis himself—"Sire, n'ayez pas peur. Le roi répondit: Je n'ai pas peur; mettez le main sur mon coeur, il est pur et tranquille"—takes place before the mob bursts in on them.[31] In *The French Revolution* the same exchange—"Sire, do not fear,' says one of his Grenadiers. 'Fear?' answers Louis: 'feel then' putting the man's hand on his heart"—takes place with "black Sansculottism weltering round him" (p. 262); and the entourage Roederer describes as accompanying Louis to his encounter with the people (the royal family, sister Elizabeth, and an array of political luminaries) recedes into shadow.[32]

Dream-grotto

Focusing on Louis XVI to the near exclusion of his supporting cast is Carlyle's method not only of posing him as an incongruity in confrontation with his (equally incongruous) opposite but of repairing if not his kingly, at least his manly image. Louis's kingly image, Carlyle (or, as Carlyle would have insisted, history) had already destroyed beyond repair: setting the royal heir against Mirabeau, in a very different sense, "a born king of men" (1: 140), who, unlike either prince of the blood to reign in his time, demonstrates himself the real heir of France's last, born king, Louis XIV. Of Mirabeau's capacity to direct events, Carlyle suggests, "he might say with the old Despot: 'The National Assembly? I am that' " (1: 137). For this capacity (metaphorically white magic as Dubarry's is black magic), Carlyle claims a power as great as the power of Dubarry. In his fight to save the throne, Mirabeau exerts over Marie Antoinette a "most legitimate sorcery" (2: 137).

Dying—consumed, as *La Révolution* is ultimately to be consumed, by his own energy, and as Louis XV was consumed by his own debauchery; burning like Hercules in a Nessus robe, and forming yet another permutation on the image of the dying king—Mirabeau sounds, even to his own ear, " 'the death-dirge of French Monarchy' " (2: 141). In a nice confluence of historical fact and allegorical logic, his preeminence in the Revolution passes to Danton, who schemes monarchy's death.

Carlyle's redemption of Louis XVI's manhood serves to distinguish between this death, of monarchy as despotism incarnate, and the death of the human being who is the monarch. It prepares us for his emergence as a tragic figure when, on January 21, 1793, *La Révolution* sends him to the guillotine. Even pickpockets, viewed sympathetically, may, Carlyle avers, "have a whole five-act Tragedy in them" (3: 107); and he ensures that Louis is viewed sympathetically on his way to the scaffold.

While Carlyle keeps to the outlines of the regicide as he found them, mainly in Mercier's *New Picture of Paris*, he suppresses the curious fascination with regicide's indignity in which Mercier indulges:

> It is indeed the same individual, crowned and consecrated at Rheims, mounted on a raised flooring, surrounded by all the nobility kneeling at his feet, hailed by a thousand acclamations, adored almost as a god, whose look, and voice, and gesture were as so many orders; he who was surrounded with honours and enjoyments, in short, separated as it were, from the rest of mankind, it is really the same man whom I see handled by four [later emended to five or six] of the executioner's helpers, and his clothes stript off with violence; whose voice is drowned by the noise of the drum, tied to a plank, still struggling, and receiving, so ill prepared, the stroke of the guillotine, that he has not his neck, but the hinder part of his head, and the jaw, horribly dissevered.[33]

In *The French Revolution*, though Carlyle depicts Louis as Mercier had, struggling with his executioners ("six of them desperate, him singly desperate"), his voice at last drowned by the drums, his final moment is caught with stark simplicity: "Abbe Edgeworth, stooping, bespeaks him: 'Son of Saint Louis, ascend to Heaven.' The Axe clanks down; a King's life is shorn away" (3: 111). Carlyle spares us the lurid details of the consequence of headsman Samson's mistiming.

He does not spare us Mercier's report of the jubilation spread throughout Paris by the axe's clank:

> His blood flows [Mercier writes]: shouts of joy from eighty thousand armed men assail my ears, and are repeated along the quays. I see the students of the Quatre Nations lifting their hats in the air. The blood flows: it is, who shall dip the end of his finger in it, a pen, a scrap of paper. One person tastes it, and says, "*It is devilishly salt.*" An executioner on the edge of the scaffold tells and distributes little packets of his hair; the rope that bound him is purchased;

everyone carries off a little fragment of his clothes, or some bloody vestige of this tragical scene. I saw the people filing off, holding by each other's arms, laughing, and talking familiarly, as if they were returning home from a fete.

There was no alteration in their countenances, and those have asserted a falsehood who stated that a stupor reigned through the city. The day of punishment made no impression: the theatres were open, the drinking-houses on the side of the bloody scene, emptied their cans as usual, and they cried cakes and petits-patés around the decapitated body.[34]

Carlyle appropriates such details as Samson's trade in locks of hair and pieces of coat and the callousness, even happiness, and indifference to consequences with which the populace regards the spectacle. He also embellishes Mercier's version with particulars drawn from other eyewitnesses: the crowd responds to Samson's display of Louis's head with a shout of *Vive la République*; D'Orleans Egalité, having watched the proceedings from his carriage, drives off; "the Townhall Councillors rub their hands, saying, 'It is done, It is done' " (3: 111).

Moreover, Carlyle removes Mercier himself as the eye and mind through which the event is filtered, and instead confronts us directly with its horror. His tactic is Conradian: to make us hear, feel, and see. Except by a cryptic note, he acknowledges his dependence on *New Picture of Paris* only when he describes France's second thoughts about Louis's death: "Not till some days after, according to Mercier, did public men see what a grave thing it was" (p. 111).

"Grave" is Carlyle's word, not Mercier's: a pun that recalls the streets preserving a "silence as of the grave" (p. 109) for the passage of Louis's tumbrel. Carlyle gives the scene the quality not of life but of myth, of nightmare: "Eighty-thousand armed men stand ranked, like armed statues of men; cannons bristle, cannoneers with match burning, but no word or movement: it is a city enchanted into silence and stone" (p. 109). Paris enchanted into silence and stone on Louis XVI's death day evokes the enchanted Dubarrydom, in which Louis XV had tried to hide from death, and foreshadows the scene, nine months later, when Marie Antoinette stands before the revolutionary tribunal: "Dim, dim, as if in disastrous eclipse; like the pale kingdoms of Dis! Plutonic Judges, Plutonic Tinville; encircled, nine times, with Styx and Lethe, with Fire-Phlegethon and Cocytus named of Lamentation! The very witnesses summoned are like Ghosts" (3: 194–95). One enchantment leads to the other, transforming Paris into a city of hell.

Carlyle-Teufelsdröckh reflects on these same paroxysms and their aftermath in *Sartor Resartus* and asks his readers, "What are all your national Wars, with their Moscow Retreats, and sanguinary hate-filled Revolutions, but the Somnambulism of uneasy Sleepers?" (p. 54). Where and when—the dimensions of history—constitute for him a Dream-grotto" (p. 55).[35] In reconstructing the executions of Louis and Marie Antoinette and the entire convulsion of the Terror, Carlyle plunges us into this dream-grotto, with the flimsy veneer of consciousness, of civilization, stripped away.

He proposes a view of the past even more radical than that of, say, Hayden White, who argues the analogousness of history and dream, or of Morse Peckham (himself a Carlyle scholar), who asks, "What do I know about the past?" and answers: "Nothing—only I know that tomorrow morning I must assume, on the evidence of the typewritten document before me, that there was a past. Every morning, unconsciously or consciously I create myself a past, and every morning, indeed every second, or that fraction of a second which is the span of mental activity, I recreate the past. And it is always different."[36] Both White and Peckham assume that the past exists, though they cannot know it, and that, therefore, attempts to represent it necessarily distort, like dreams. To the writer of *Sartor Resartus* the past itself, rather than the attempt to represent it, is the dream. As Teufelsdröckh declares (in a scrap of wisdom that would strongly impress Hardy), God, "the Unslumbering, whose work both Dream and Dreamer are, we see not" (p. 53).

Because we see not God, those who claim rigor for any particular conception of history mislead: "They only are wise," Teufelsdröckh adds, in a pronouncement Peckham may have been remembering as he framed his fluid conception of history, "who know that they know nothing" (p. 54). Or, as Yeats toward the end of his life wrote to Lady Elizabeth Pelham, "Man can embody truth but he cannot know it."[37] To embody truth is, White suggests, Carlyle's prescription for writing history.

White numbers Carlyle among those he labels ideographic historians: historians for whom historical narration consists in weaving a fabric of images.[38] *The French Revolution* is a fabric of images that add up to a mammoth psychic image. Both Albert J. LaValley and Philip Rosenberg observe that Carlyle hypostatizes the unconscious, acknowledging the irrational, the demonic, as a force in history.[39] Developing the thesis that the Revolution took place first in the minds of Frenchmen, he makes his narrative an exploration of the French mind. If, as Rosenberg especially stresses, Carlyle's metaphysics looks back to Blake, his psychology looks forward to Freud.[40]

He indicts the philosophes for, among other things, their excessive confidence in human reason, and he asks the spokesmen for change (of whom the philosophes were the chief in prerevolutionary France) whether they have "well considered all that Habit does in this life of ours; how all Knowledge and all Practice hang wondrous over infinite abysses of the Unknown, Impracticable; and our whole being is an infinite abyss, *overarched* by Habit, as by a thin Earth-rind laboriously built together" (1: 38). In this question Carlyle sums up the dynamics of his French Revolution. The urgency of his tone, the warning it contains, stem from his belief that "every man [not just every Frenchman] . . . holds confined within him a *mad*-man" (p. 38) who, freed from the restraints of habit, would destroy society. Carlyle's madman easily translates into Freud's id, his concept of habit into the superego. What happens when habit loses its power to repress the madman within every man is epitomized by "Mad Paris . . . abandoned altogether to itself" two days before the storming of the Bastille:

> What a Paris, when the darkness fell! A European metropolitan City hurled suddenly forth from its old combinations and arrangements; to crash tumultuously together, seeking new. Use and want will now no longer direct any man; each man, with what of originality he has, must begin thinking; or following those that think. Seven hundred thousand individuals, on the sudden, finding all their old paths, old ways of acting and deciding vanish from under

their feet. And so there go they, with clangour and terror: from above, Broglie, the war-god, impends, preternatural, with his redhot cannon-balls; and from below a preternatural Brigand world menaces with dirk and firebrand: madness rules the hour. (1: 178–79)

"Above" and "below" reiterate the metaphor of earth-rind covering the abyss as habit confines madness, a metaphor that acquires volcanic implications when Carlyle (exploring the restiveness of prerevolutionary France) asks, through which of "so many cracks [issuing sulphur smoke] . . . will the main Explosion carry itself?" (1: 61). The personification of "above" and "below" in Broglie and the brigands also introduces the allegory Carlyle will evolve as a vehicle for depicting the psyche of France out of control.

Broglie and his commander on the scene, Besenval, helpless before Paris's effervescent citizenry, foreshadow Bouillé at Nancy who, stepping aside to consult with Inspector Malseigne, loses his precarious grip on the Sansculottic mob; or Danton who, wearied by the Terror, withdraws to Arcis to recuperate and loses his precarious grip on the Revolution itself. They embody authority, the restraints society imposes on the madman within; when their vigilance relaxes, the madman breaks loose.

The madman assumes form in the gambolling, destructive giant, Sansculottism, and his individual human counterparts, mainly Robespierre and Marat. When, on ninth Thermidor (July 27, 1794), Sansculottism "suicidally 'fractured its underjaw' and lies writhing, never to rise more" (3: 288), the giant and Robespierre coalesce. How the giant comes to this end is suggested by the rumors (already current in 1789) of brigands who menace the countryside with dirk and firebrand. For these rumors prefigure Marat's own terroristic plan—"two hundred Naples Bravoes, armed each with a good dirk, and a muff on his left arm by way of shield" (2: 16)—for completing the Revolution.

Proposed in the atmosphere of ostensibly universal good will to be celebrated at the Feast of Pikes, the plan is belittled, its designer satirized as Cassandra-Marat. But, Carlyle asks portentously, "Were it not singular if this dirk-and-muff plan of his (with superficial modifications) proved to be precisely the plan adopted?" (2: 16). Carlyle's question implies the process—a conception in the mind emerging as a concretion in the world—whereby violence overspreads France. The process is foreshadowed by Besenval, at five on the morning of the Bastille, dreaming of a mysterious but (apparently) live product of the streets who warns, also Cassandra-like, "that if blood flowed woe to him who shed it" (1: 187). As this warning, impressed on the half-awake Besenval's consciousness, is translated into action in the slaughter of De Launay after his surrender of the Bastille, so the plan devised by Cassandra-Marat is translated into action, initially, in the massacre of the Swiss during the Insurrection of August 10, 1792. Thus (in an episode Carlyle seems to have invented) Marat is portrayed pulling the town hall bell on August 9–10 to rally Sanscullottic Paris, while "Robespierre lies deep, invisible for the next forty hours" (2: 20).

The voice urging murder has risen from "subterranean" Paris to repress the voice (in the end equally murderous) urging policy and caution. Los calling his sons to the strife of blood has, briefly, routed Urizen. In the structural scheme of Carlyle's narrative the last book of Part II, "Constitution Burst in Pieces," is followed by Part III, "The Guillotine."

The guillotine is the heart of the Revolution: "The clanking of its huge axe, rising and falling there, in horrid systole-diastole, is portion of the whole enormous Life-movement and pulsation of the Sansculottic System" (3: 193). *La Révolution* materializes as a living being, the protagonist in a kind of monstrous psychomachia.[41] In allegory, as Angus Fletcher points out, the hero is less an integrated character than a generator of fragments, aspects of his inner self. Aspects of *La Révolution*'s inner self describe, in Carlyle's rendering, the men generally thought by history to be her major architects. Fletcher attributes to the allegorical hero conduct akin to that of men possessed by daemons.[42] From Lafayette, "fast anchored to the Washington Formula" (2: 308), to the Marseillais, who know how to die, to Napoleon delivering his whiff of grapeshot, all of Carlyle's chief personages behave as men possessed. They embody fixed ideas.

Marat, labeled a "fanatic Anchorite" (2: 16) and consumed

with desire for revenge, epitomizes the fixed idea as energizing revolutionary force. In the "Improvised Commune" of Septembrists, that assumes power after August 10 and that initiates the Terror, he—no longer crouched in Legendre's cellar—rises to "a seat of honour" (3: 8). At the fall of the Girondins on June 2, 1793 (Year One of Liberty, Equality, Fraternity), he expresses the Sansculottic will, ordering "us, in the Sovereign's name, to return to our place, and do as we are bidden and bound" (3: 162).

Carlyle reinforces the symbolic correspondences he incorporates in historical figures by attaching identifying images to them. The image identifying Marat is that of a mad dog. As Carlyle, reflecting on Marat seated among the Septembrists in his *tribune particulière* remarks, "All dogs have their day; even rabid dogs" (3: 8). Marat's day reaches its zenith in his triumph over the Girondins. When he orders the Convention to return to its place, and "the Convention returns" (3: 162), he fulfills the role announced for him in the "Improvised Commune," as *"Keeper . . .* of the Sovereign's Conscience" (3: 8).

That Carlyle means Marat to be understood not as a mere aberration but as the realization of savage impulses deeply rooted in the human psyche is established by the bestial imagery he employs to characterize violent France before Marat's emergence as a power. The forty-foot-high gallows, with which the king suppresses the Sansculottes' clamor for bread on May 2, 1775, drives the clamorers "back to their dens,—for a time" (1: 34). The destruction on April 27–28, 1789, of Le Sieur Réveillon's warehouse—for the manufacture of paper, Carlyle is careful to tell us—heralds the rise of an as yet unknown political evangel "in those dark dens, in those dark heads and hungry hearts" (1: 128). The Place de Grève, where after the storming of the Bastille, one of De Launay's *invalides* is hanged and De Launay himself beheaded, "is become a Throat of the Tiger" (1: 197). The rebellious Mestre-de-Camp at Nancy chases Inspector Malseigne as Diana's hounds chased Actaeon (2: 90).

Marat risen from "subterranean" Paris is, then, the natural outgrowth, the ultimate materialization, of a rage fueled by centuries of abuse. When the Girondins emerge from the Tuileries to find him blocking their path, they confront a monster partly of their own making. Though they are heirs of the philosophes in their belief that the republic can and should be founded on reason, they have cultivated the violent revolution Marat embodies, seeking to turn it to their own use. Grown beyond their strength to control, it consumes them—literally. Buzot and Petion, the last surviving Girondins, declared "out of law," and in flight, are found dead in a cornfield, "their bodies half-eaten by dogs" (3: 201).

The image of the rabid dog, like Blake's kings consuming and ultimately consumed, reiterates Carlyle's portrayal of the Revolution as consumed by its own energy. Marat dies at the hands of Charlotte Corday. The Girondins die at the hands of the Jacobins. The Jacobins die at the hands of each other. Philip Rosenberg stresses the ambivalence with which Carlyle viewed this bloody orgy, arguing that he wished to see the Revolution completed without having to see it repeated.[43] Although no one, I think, would be more surprised than Carlyle himself at Rosenberg's detection of an incipient Marxist in the author of *Sartor Resartus* and *The French Revolution*, Rosenberg is surely right to insist upon the Carlyle who fulminated at social injustice, whether in prerevolutionary France or in (potentially prerevolutionary?) England. It is as an outburst against injustice that Carlyle would have us read what he describes in *The French Revolution* as the shout greeting Robespierre's end, prolonged "over Europe, and down to this generation" (3: 285).

For *The French Revolution* is addressed to "this generation": "That there be no second Sansculottism in our Earth for a thousand years, let us understand well what the first was" (3: 313). The final book of the narrative thus joins a historical conclusion to a personal coda that seeks to defuse England's indignation at French excesses. "There is," Carlyle observes, "no period . . . in which the general Twenty-five Millions of France suffered *less* than in this period . . . they name Reign of Terror" (3: 312), and he reminds his countrymen (prophetically) that England breeds its own *enragés*: "The Irish Sans-potato, had he not senses then, nay a soul!" (p. 312). The madman hunger awakens in France, and who eventuates in Marat, lurks in every man; the explosive heat of revolution simmers beneath the earth-rind of every society. Cagliostro's prophecy is not simply "fulfilled" (past tense); it is still "fulfilling."

The French Revolution closes, therefore, not with Cagliostro prophesying, nor with Carlyle the historian historicizing, but with Carlyle the preacher delivering a brief homily on the reciprocal obligations of writers and readers. If history comprises "the letter of Instructions, which the old generations write and posthumously transmit to the new" (*CME* 3: 167), the historian serves as scribe to those old generations. What is at stake in the effort to provide a faithful transcription of their letter of instructions is, finally, not the past but the future. Given a stake of such importance, Carlyle can legitimately warn, when he bids farewell to the companion of this arduous journey through space-time, "Ill stands it with me if I have spoken falsely: thine also it was to hear truly" (3: 323).

FIVE

Antihistory: Dickens' *A Tale of Two Cities*

Correspondent for Posterity

In an essay published in *The National Review* a year before *A Tale of Two Cities* (1859) appeared, Walter Bagehot likens Dickens describing one of those cities to "a special correspondent for posterity." Bagehot finds in Dickens a newspaperman's eye for London life: "His memory is full of instances of old buildings and curious people, and [because, as Bagehot sums up London, 'everything is there, and everything is disconnected'] he does not care to piece them together . . . each scene . . . is a separate scene,—each street a separate street." Dickens, Bagehot concludes, is gifted with "the peculiar alertness of observation . . . observable in those who live by it."[1]

Bagehot also suggests that though Dickens does not attempt to digest an array of histories, memoirs, and journalistic ephemera to comprehend the past (as Carlyle does), he nonetheless ascribes to himself writing fiction the motive that Carlyle ascribes to the historian writing history: to transmit the old generation's posthumous letter of instructions to the new. And in *A Tale of Two Cities* he appropriates aspects of Carlyle's narrative method. As Carlyle, through personal pronouns and present tense, conveys his reader into the Paris of 1774–1795, so Dickens, if sporadically, employs the same devices to the same end, encompassing almost the same span of years.[2] His technique, even when he has not in fact been there, partakes of the correspondent's peculiar alertness of observation.

"I have," as he puts it in his preface to *A Tale of Two Cities*, "so far verified what is done and suffered in these pages, as that I have certainly done and suffered it all myself" (p. 29).

He essentially repeats this claim in the novel, suspending the narrative to generalize from the conduct of Charles Darnay's fellow prisoners in La Force that, "In seasons of pestilence, some of us will have a secret attraction to the disease. . . . And all of us have like wonders hidden in our breasts, only needing circumstances to evoke them" (p. 310). "Us" invites the reader to perceive in this brief excursus Dickens not only arguing the psychological validity of his characters—Darnay and Sydney Carton are, each in his own way, attracted to the disease—but also reflecting on himself. Numerous critics have speculated on the degree to which Dickens' unhappy marriage and his performance opposite Ellen Ternan in *The Frozen Deep* underlie the triangle of Lucie Manette, Darnay, and Carton. In giving his own initials to the protagonist who survives, and associating himself with the inmates of La Force, Dickens seems, however, to imply something more: that, having read Carlyle's *French Revolution* (five hundred times, he said) and visited the scenes Carlyle wrote about, as well as having witnessed, from a safe distance, the Revolution's aftershocks in 1830 and 1848, he was fascinated, his imagination caught by the upheaval. The novel manifests his response to a power that draws him, as Darnay is drawn, "to the Loadstone Rock."

Dickens can come to the Loadstone Rock only vicariously, through the experience of his characters. The incisiveness he achieves in exploiting both methods of apprehension—as a journalist observing, as a novelist inventing—is what Bagehot particularly praises. Dickens, who may well have read Bagehot's essay, alerts the reader to the element of journalistic observation in his technique of novel writing almost before *A Tale of Two Cities* starts to unfold: "A wonderful fact to reflect upon, that every human creature is constituted to be that profound secret and mystery to every other. A solemn consideration, when I enter a great city by night, that every one of those darkly clustered houses encloses its own secret; that every beating heart in the hundreds of thousands of breasts there, is, in some of its imaginings, a secret to the heart nearest it" (p. 44).

In Bagehot's newspaperman's eye view of London, "There is every kind of person in some houses; but there is no more connection between the houses than between the neighbours in the list of 'births, marriages, and deaths.' "[3] This lack of connection, even among neighbors in houses, is epitomized in *A Tale of Two Cities* by the isolation of Mr. Lorry and his fellow passengers on the Dover mail: "All three were wrapped to the cheek-bones and over the ears, and wore jack-boots. Not one of the three could have said, from anything he saw, what either of the other two was like; and each was hidden under almost as many wrappers from the eyes of the mind, as from the eyes of the body, of his two companions" (p. 38). Mr. Lorry's companions leave the coach before Dover and disappear from the novel. But Mr. Lorry himself engenders a mystery which, alluding to, yet masking, the reason for his journey, forcefully involves the reader. When Jerry Cruncher arrives with the message, "Wait at Dover for Mam'selle," and receives the cryptic answer, "RECALLED TO LIFE" (p. 41), the perplexity he, the guard, and the driver feel is a perplexity we share:

"What do you make of it, Tom?"
"Nothing at all, Joe."
"That's a coincidence, too," the guard mused, "for I made the same of it myself."

(p. 42)

Making Something of It

Making nothing of it is, Taylor Stoehr argues, a position into which Dickens characteristically maneuvers his readers in his opening episodes.[4] Stoehr takes this device to be part of Dickens' strategy for rendering his fiction dreamlike. Although Stoehr's proposal seems to me to account for the disjunctions in the plots of many (perhaps most) Dickens novels, it fails adequately to address the epistemological concerns of *A Tale of Two Cities*, how to "make something" of experience. The point of view Dickens establishes for his narrator in this novel moves, with the flexibility of a camera, from the first chapter's panorama of the years between 1775 and his own time to the second chapter's drastically narrowed focus on one night in one month of one year along one road, and the apparent non-sequitur of which Jerry, the guard, and driver can make nothing.

The titles of Chapters 1 and 2—"The Period" and "The Mail"—tell something of the tale behind the *Tale*. To survey the period suggests the historian's practice of standing outside events and trying to see them whole. To dramatize the coach journey of a single character locked among strangers within walls of mutual suspicion exemplifies the practice of the novelist, who projects his reader into a consciousness engaged not in rationalizing history but, moment by moment, in living it. That Mr. Lorry's grasp of his mission is only less tenuous than Jerry's is stressed by the ghostly Dr. Manette he recurrently dreams as the coach rolls toward Dover:

> A hundred times the dozing passenger inquired of this spectre:
> "Buried how long?"
> The answer was always the same. "Almost eighteen years."
> "You had abandoned all hope of being dug out?"
> "Long ago."
> "You know that you are recalled to life?"
> "They tell me so."
> "I hope you care to live?"
> "I can't say."
> "Shall I show her to you? Will you come and see her?"
>
> (pp. 46–47)

Dr. Manette's inability, in Lorry's dream, to say whether he cares to live, or to see his daughter, is Lorry's inability to say what digging Dr. Manette out will bring.

Lorry's perspective and that of the reader—as yet even more surrounded by "night shadows" than Lorry—differ radically from the perspective of the narrator who, with ironic relish, catalogues the parallels in political and social circumstance between the mid-1770s and mid-1850s. This narrator enjoys the breadth of vision of Blake's Bard, who sees present, past, and future; or of Carlyle's historian who, perched on some coign of vantage, presciently looks down on the procession of the Estates-General; or of Hardy's Phantoms who, situated in the Overworld, as in a theater box, watch the Immanent Will work out its convulsive designs. He manifests Dickens assuming, in Jonathan Arac's phrase, the role of "commissioned spirit," conveying the reader to a height from which he can see reality whole.[5]

But neither Dickens, Carlyle, nor Hardy is content to view history from afar. They share Rudolf Bultmann's insistence that history becomes meaningful only through the historian's involvement.[6] Bultmann claims a Christian schema for history. Dickens, following Carlyle, also detects a Christian schema, even in the apparent anarchy of the French Revolution; and he makes his narrator a prophet who discovers order in that anarchy. The balanced antitheses in the narrator's preamble to his story—"It was the best of times, it was the worst of times, it was the age of wisdom, it was the age of foolishness" (p. 35)— suggest the voice of the Preacher in Ecclesiastes, who "gave [his] heart to know wisdom, and to know madness and folly" (1:17).

To learn that "the period [of *A Tale of Two Cities*] was so far like the present period, that some of its noisiest authorities insisted on being received, for good or for evil, in the superlative degree of comparison only" (p. 35) is, for Dickens' narrator, to learn, as the Preacher learns, that "there is no new thing under the sun" (1:9). And both employ the same method of learning. They not merely observe history, they immerse themselves in it. "I sought in mine heart," the Preacher reports of his experiment with one excess of human conduct, "to give myself unto wine, yet acquainting mine heart with wisdom; and to lay hold on folly, till I might see what was that good for the sons of men, which they should do under heaven all the days of their life" (2:3). He gains wisdom by experiencing folly, as Dickens' narrator discovers the attraction of Frenchmen to the disease by experiencing their terror. The disembodied but pervasive presence of the narrator is Dickens' vehicle for coming to the Loadstone Rock. He shares the anxiety of flight from the disease by joining Mr. Lorry, Dr. Manette, and the Darnays in their coach as they slip out of Paris. He shares Carton's religious transport at repeating Christ's sacrifice by merging his consciousness with Carton's consciousness at the scaffold.

Overcoming History

That Carton vests his survival in the memories of the Darnays and their offspring limits the extent to which Christian eschatology serves Dickens as an ordering principle. The Preacher of Ecclesiastes also diverged from orthodoxy; and it was probably only the postscript ascribed to a disciple—"Fear God, and keep his commandments: for this is the whole duty of man" (12:13)—that persuaded the rabbis to include his book in the canon. Carton, whose forays into dissolute living (we are surely to suppose) outstrip the Preacher's, comes to God by the same path. He passes through folly to a wisdom summed up in keeping God's commandments, acknowledging Christ as "the Resurrection and the Life." The narrator's participation in Carton's consciousness allows him to transcend the stance of historian and emerge as prophet. While in "The Period" he had, like Scott in *Waverley*, looked back on events from "sixty [or more] years since," in "The Footsteps Die out For Ever" (the novel's final chapter) he stands with Carton and looks forward. Indeed, he echoes Carlyle's Cagliostro at the end of *The French Revolution*, reciting his roll of victims—the king, queen, Iscariot Egalité, De Launay—consumed by the flames: "I see Barsad, and Cly, Defarge, the Vengeance, the Juryman, the Judge, long ranks of new oppressors who have risen on the destruction of the old, perishing by this retributive instrument" (p. 404).

The point of view of Cagliostro-Carlyle is entirely retrospective. The point of view of Carton-Dickens becomes prospective in Carton's forecast of a France redeemed within history: "I see a beautiful city and a brilliant people rising from this abyss, and, in their struggles to be truly free, in their triumphs and defeats, through long years to come, I see the evil of this time and of the previous time of which this is the natural birth, gradually making expiation for itself and wearing out" (p. 404).

Though Robert Alter attributes the strength of *A Tale of Two Cities* to its recognition of the power of evil in history, he pronounces the novel flawed by its rhetorical weighting toward the optimism of Carton's prophecy. Dickens, he argues, "tries hard, through the selfless devotion of his more exemplary characters, to suggest something of mankind's potential for moral regeneration; but he is considerably less convincing in this effort, partly because history . . . offers so little evidence which the imagination of hope can use to sustain itself."[7] Alter's judgment says more about his ideological frame of reference, his philosophy of history, than about Dickens' novel. If we reject the Christian confidence that (in words Joyce has Mr. Deasy murmur to Stephen), "All history moves toward one great goal, the manifestation of God," we may well find ourselves unmoved by Dickens' happy ending.

If we also reject Marx's confidence that history moves inexorably toward revolution, we will, with George Orwell, recognize the flimsiness of T. A. Jackson's case for Dickens as a revolutionary.[8] One of the most influential Marxist critics of the 1970s and 1980s, Fredric Jameson, defines history, in proper Marxist fashion, as "the experience of Necessity": not the content but the "*form* of events," a rigorous narrative category that "refuses desire and sets inexorable limits to individual as well as collective praxis."[9] Taken seriously as a historical novel, *A Tale of Two Cities* challenges this definition. For Carton and the Darnays, history yields to desire. Dickens even allows Carton to perceive in death an escape from history. As he awaits the ax, he comforts the frail seamstress, worried lest the time to pass before her beloved cousin joins her in "the better land" seem long, assuring her that "It cannot be, my child; there is no Time there, and no trouble there" (p. 403).

Carton's visions, of a land with no time and no trouble, and of Paris as a beautiful city filled with brilliant people, are foreshadowed by his vision on leaving Stryver's flat after a night of playing jackal to his employer's lion: "In the fair city of this vision, there were airy galleries from which the loves and graces looked upon him, gardens in which the fruits of life hung ripening, waters of Hope that sparkled in his sight" (pp. 121–22). Carton's fair city is, he presumes, to be found nowhere in this world. Yet it exists in miniature at the corner of Soho occupied by Lucie and Dr. Manette: "The summer light struck into the corner brilliantly in the earlier part of the day; but, when the streets grew hot, the corner was in shadow, though not in shadow so remote but that you could see beyond it into the glare of brightness. It was a cool spot, staid but cheerful, a wonderful place for echoes, and a very harbour from the raging streets" (p. 123).

Their home is Edenic, seemingly insulated from the city around it and from the city across the Channel. It has the idyllic quality that Frank Kermode emphasizes as peculiar to the *aevum*, a locale participating in both time and eternity without belonging to either, where "things can be perpetual without being eternal."[10] Dr. Manette, Lucie, and Charles live in such a locale. Though he has been immured in the Bastille for eighteen years, Dr. Manette regains his intellect and his vigor; after ten of those years, he can still—insisting the while on its meticulous accuracy—record the indictment of the Evrémondes that Defarge finds in his cell. Though Lucie and Charles age twenty years from beginning to end, they never appear any older.

Dickens' success in impressing their perpetually youthful images on the minds of his readers is illustrated by George Woodcock's description of *A Tale of Two Cities* as a story where "one of two young men sacrifices himself so that the other may enjoy happiness with the girl whom both of them love"—terms hardly appropriate to a situation in which, however old Carton may be when he goes to the guillotine, Charles is forty and Lucie thirty-seven.[11] We might, ordinarily, dismiss this matter as the kind of license we grant the novelist. Nineteenth-century literature is rife with bad romances, often imitative of Scott or Dickens, in which heroes and heroines seem arrested in youth. Woodcock reads in Lucie's immunity to time a Dickensian device for portraying her as, in spirit, always a maiden because "romantically the maiden is associated with death."[12]

But *A Tale of Two Cities* draws attention to Lucie's youth as, more than a conventional motif, a phenomenon that borders on the preternatural, because the novel repeatedly focuses on the years passing. Lucie at age seventeen accompanies Mr. Lorry to Paris to rescue her father in 1775. She and Dr. Manette testify at Charles's trial for treason in 1780, when he is twenty-five. Charles declares his love for Lucie in 1781. Between 1781 and 1789, they marry and have two children, one of whom dies. At the Revolution's outbreak, Charles is thirty-four, Lucie thirty-one. Charles returns to France at the plea of his retainer, Gabelle, in 1792. The Darnays, Dr. Manette, Mr. Lorry, and their servants, Jerry Cruncher and Miss Pross, finally escape from France in 1795.

Time-spaces

This sequence of dates—all carefully specified, as if Dickens had plotted the action on a calendar—suggests that two temporal dimensions coexist in the novel: the *aevum*, in which duration changes nothing; and history, in which duration changes everything. They intersect through Charles's receipt of Gabelle's letter, which recalls him, with his family and friends, to the world of history.

The novel's temporal dimensions are thus tied to the two cities of its title. Not that London becomes for Dickens a city exempt from history. Though the Darnays, Dr. Manette, even Carton, Mr. Lorry, and Miss Pross try assiduously to ignore the echoes and rumblings in the raging streets, Lucie and Carton can sense the threat the streets pose to their domestic idyll. Carton's vision of the fair city takes place in a London perceived as "wilderness." That his vision is characterized as "a mirage of honourable ambition, self-denial, and perseverance" (p. 121) identifies it as a mental state distinct from his outward show, and buried inside him. Lucie hears in the echoes that penetrate their haven footsteps "coming by-and-by into our lives" (p. 133). Carton's awareness of moral values he would uphold despite himself, and on which he will someday act by offering himself as a sacrifice to those raging streets, and Lucie's awareness of an as yet indefinable force to shatter her tranquillity and thrust her, with family and friends, into those raging streets imply a correlation between the public struggle in the streets and the private struggles they wage: Carton to keep the savagery of the streets from overwhelming his humanity, Lucie to keep that savagery (the impact of which she has already felt in her father's ordeal) from invading their lives.

Both concede, at least subconsciously, that they cannot forever insulate themselves or their loved ones from history. The novel's two cities configure a psychic allegory. Lucie in Soho, "busily winding the golden thread which bound her husband, and her father, and herself, and her old directress and companion, in a life of quiet bliss" (p. 239), counterbalances Madame Defarge in Saint-Antoine, busily knitting the record that binds her victims to a grim reality and targets them for revolutionary vengeance.[13]

Though Madame Defarge is, in the end, the novel's irredeemable villain, the epitome of evil, in contrast to Lucie, its heroine and paragon of virtue; and though the fight between Madame Defarge and Miss Pross is partly meant, as Alter points out, to symbolize the conflict between French bestiality and English humanity, *A Tale of Two Cities* should not be reduced to Dickens' concurrence in, say, Burke's indictment of the Revolution.[14] As most Dickens critics acknowledge, his attitude is expressed by the warning embedded in the summary statement that opens "The Footsteps Die out For Ever": "Crush humanity out of shape once more, under similar hammers, and it will twist itself into the same tortured forms. Sow the same seed of rapacious license and oppression over again, and it will surely yield the same fruit according to its kind" (p. 399).

Dickens is echoing Carlyle, who makes Hosea's prophecy, "They have sown the wind, and they shall reap the whirlwind" (8:7), a refrain in his *French Revolution*. Madame Defarge, who admonishes her husband, when he pleads for mercy toward Darnay's family, to "tell Wind and Fire where to stop . . . but don't tell me" (p. 370), is the whirlwind reaped by the Evrémondes through their rape of her sister and murder of her brother. She embodies Dickens' principle that humanity oppressed will assume savage shapes: "imbued from her childhood with a brooding sense of wrong, and an inveterate hatred of a class, opportunity had developed her into a tigress" (p. 391).

Stout, strong-featured, and dark, whereas Lucie is (literally) light and delicate, Madame Defarge plays witch to Lucie's angel in the fairy tale or Gothic romance that unfolds within history in this novel. Transformed into a beast by the "terrible enchantment" (p. 263) of the Evrémondes, the old regime they represent, and the revolution it provokes, Madame Defarge imposes an equally terrible enchantment on the Vengeance, Jacques Three, and the Road-mender turned Wood-sawyer. Lucie also enchants, but to a different end, for she has "the power of charming [the] black brooding from [her father's] mind" (p. 110).

Lucie is a type of Rapunzel, reaching out with her long blond hair to rescue her father from his nightmares: "She was the golden thread that united him to a Past beyond his misery, and to a Present beyond his misery" (p. 110). She makes of the golden thread a talisman to protect her corner of Soho against the vicious world. While in Paris "three years of tempest [i.e., 1789 to 1792, are] consumed," in London "three more birthdays of little Lucie [are] woven by the golden thread into the peaceful tissue of the life of her home" (p. 263). And as the white magic that Lucie weaves with her golden thread into the tissue of life in her home suspends, if not the fact of time, then its effect, the black magic that Madame Defarge knits with her death sentences into the tissue of the Revolution suspends consciousness of time, violently contracting events to intensify their effect:

> There was no pause, no pity, no peace, no interval of relenting rest, no measurement of time. Though days and nights circled as regularly as when time was young, and the evening and the morning were the first day, other count of time there was none. Hold of it was lost in the raging fever of a nation, as it is in the fever of one patient. Now, breaking the unnatural silence of a whole city, the executioner showed the people the head of the king—and now, it seemed almost in the same breath, the head of his fair wife which had had eight weary months of imprisoned widowhood and misery, to turn it grey. (p. 302)

Dickens again echoes Carlyle, primarily his portrait of Paris during the nightmare of the king's execution. But Dickens does not, as several of his critics suggest, slavishly follow Carlyle.[15] Thrusting his characters into "the deluge of the Year One of Liberty" (p. 301), Dickens instead explores (what Carlyle only implies) the delirium of acting on the belief that a nation, any more than an individual, can jettison the past and begin anew as if every evening and every morning were the first day.

That Lucie, her family, and friends try to live such a perpetual beginning in Soho, as Madame Defarge, her husband, and their comrades try to live their perpetual beginning in Saint-Antoine, accounts for the dreamlike atmosphere Stoehr detects in *A Tale of Two Cities*. Stoehr observes, perhaps according Dickens greater psychological awareness than a pre-Freudian could have been expected to possess, that fiction imitates not the dream but the dreamer's telling of the dream, which is our only access to it. The hero in fiction is the focus of a "secondary elaboration": the narrator's attempt, analogous to the collaborative efforts of dreamer and analyst, to make what the conscious self would call realistic sense out of nightmare or hallucination.[16]

The narrator, free to maintain whatever distance he chooses from the action, and not the hero, always embroiled in the action, thus remains Dickens' central intelligence throughout. Released after hearing the jury in the Old Bailey pronounce him innocent, Charles is as disoriented as Paris is by the tide of revolution. When Carton asks Charles, over dinner, whether he feels that he belongs "to this terrestrial sphere again," Charles answers, "I am frightfully confused regarding time and place; but I am so far mended as to feel that" (p. 114).

Confusion about time and place also afflicts Dickens' Parisians who, like Carlyle's, reenact Genesis in their streets, on the Champ-de-Mars, or at the Place de la Révolution. Instead of seeking an explanation of how time enters experience as they do, in the King James version of the First Day, the narrator seeks his explanation (filtered, it may be, through myriads of intervening minds) in Heraclitus. The fountain by which the Marquis St. Evrémonde's carriage runs down Gaspard's child metamorphoses into a symbol of time as a flowing stream: "The water of the fountain ran, the swift river ran, the day ran into evening, so much life in the city ran into death according to rule, time and tide waited for no man, the rats were sleeping close together in their dark holes again, the Fancy Ball was lighted up at supper, all things ran their course" (p. 143).

For the reader, who can look back on the Revolution complete, the fountain foreshadows the revenge to overtake not alone the Marquis but his class. In the plot of the novel—the vehicle for Dickens' insistence that, among characters advertently or inadvertently immersed in history, a correlation exists between individual and collective fate—the fountain also prefigures two other fountains, one in the village below the Evrémondes' chateau, the other at the chateau. Both count down the hours to Gaspard's personal revenge against the Marquis: "The fountain in the village flowed unseen and unheard, and the fountain at the chateau dropped unseen and unheard—both melting away, like the minutes that were falling from the spring of Time—through three dark hours" (p. 157).

And the Marquis is not the only character whose last hours are perceived to pass like water flowing. When, moments before she is killed by her own pistol, Madame Defarge surprises Miss Pross into spilling her basin, the narrator remarks on the "strange stern ways, and . . . much staining blood" by which Madame Defarge's "feet had come to meet that water" (p. 394). When the fifty-two victims, of whom Carton is one, board the tumbrels for their ride to the Place de la Révolution, they are said "to roll . . . on the life-tide of the city to the boundless everlasting sea" (p. 375). When Carton places his neck in the guillotine, the impressions that crowd in upon him before the ax are of "the murmuring of many voices, the upturning of many faces, the pressing on of many footsteps in the outskirts of the crowd, so that it swells forward in a mass [until], like one great heave of water, all flashes away" (p. 403).

Dickens distinguishes between what Hans Meyerhoff, borrowing from Bergson, labels *le temps humain*—time as a continuous flow—and the physicists "spatialization" of time, its division into discrete, quantified units.[17] The lives of Dr. Manette, Lucie, and Charles in Soho—seemingly static stretches bounded by events that occur on particular dates and in particular places—suggest spatialized time. The lives of the poor in Saint-Antoine or in the village below the Evrémondes' chateau—crushed into pathetic shapes by a historically sanctioned oppression—epitomize *le temps humain* in prerevolutionary France, borne along on time's current toward 1789. Among the witnesses to the altercation at the fountain between Defarge, speaking for Gaspard, and the Marquis is Madame Defarge, "knitting . . . with the steadfastness of Fate" (p. 143); and Madame Defarge, encoding in her handiwork the tide to overtake the Marquis and his class, foreshadows her cohorts—meant, as Alter points out, to evoke the Fates—knitting before the guillotine and counting the heads as they fall: One . . . Two . . .

until finally, the blade crashing down on Carton, Twenty-Three.[18]

The Fitness of Things

Beneath the aura of ritualistic formality these myth-like figures give to Carton's execution is Dickens' sense, not to be shrugged off as mere fantasy, of how it must have been. For he had witnessed just such an execution in Rome, probably in 1845, and recorded it in lurid detail:

> [The condemned man] immediately kneeled down, below the knife. His neck, fitting into a hole, made for the purpose, in a cross plank, was shut down, by another plank above; exactly like a pillory. Immediately below him was a leathern bag. And into it his head rolled instantly.
>
> The executioner was holding it by the hair, and walking with it round the scaffold, showing it to the people, before one quite knew that the knife had fallen heavily, and with a rattling sound.[19]

The knife falling heavily with a rattling sound (a detail that escaped Carlyle) can be heard in *A Tale of Two Cities* in the "Crash!" punctuated by the toll of the knitting women.

For Carlyle's reliance on his ability to extrapolate reality from the printed word, Dickens substituted his ability to build analogies from personal observations, translating them into the world he sought to imagine into being. His visit on March 8, 1842, to the then famous Cherry Hill Prison in Philadelphia, and his impression of the prisoners as men "buried alive; to be dug out in the slow round of years," thus lurked in his memory for over a decade, to be recovered as a datum in constructing the ordeal of Dr. Manette.[20]

Realism is, indeed, a quality Dickens insistently claimed for *A Tale of Two Cities*. In a letter to Sir Edward Bulwer-Lytton, written June 5, 1860, he answers the objection that the right granting the Evrémondes sexual privilege with Madame Defarge's sister had, by 1757 (the date of their crime), already been abolished: "I see no reason to doubt, but on the contrary, many reasons to believe, that some of these privileges had been used to the frightful oppression of the peasant, quite as near the time of the Revolution as the doctor's narrative, which, you will remember, dates long before the Terror. And surely when the new philosophy was the talk of the salons and the slang of the hour, it is not unreasonable or unallowable to suppose a nobleman wedded to the old cruel ideas, and representing the time going out, as his nephew represents the time coming in." Dickens asserts not the accuracy but the plausibility of the situation. In the same letter he defends the struggle he wholly invents, between Madame Defarge and Miss Pross, as also historically, psychologically, plausible:

> I am not clear, and I never have been clear, respecting that canon of fiction which forbids the interposition of accident in such a case as Madame Defarge's death. Where the accident is inseparable from the passion and emotion of the character, where it is strictly consistent with the whole design, and arises out of some culminating proceeding on the part of the character which the whole story has led up to, it seems to me to become, as it were, an act of divine justice. And when I use Miss Pross (though this is quite another question) to bring about that catastrophe, I have the positive intention of making that half-comic intervention a part of the desperate woman's failure, and of opposing that mean death—instead of a desperate one in the streets, which she wouldn't have minded—to the dignity of Carton's wrong or right; this was the design, and seemed to be in the fitness of things.[21]

A design that affirms the fitness of things hints, Dickens implies, at a history providentially ordered, shaped "as it were" by divine justice. *A Tale of Two Cities* contains in microcosm what he, like Blake and Carlyle, thought to be universal history. Dickens' French Revolution too carries suggestions of an impending apocalypse. The mob that breaks down the walls of the Bastille is compared, in a metaphor evoking the Flood but probably suggested by Carlyle, to "the sea that rushed in, as if there were an eternity of people, as well as of time and space" (p. 246).[22] The Revolution becomes, in the eyes of its hostile beholders, Monseigneur in exile and "British orthodoxy" according to Burke, "the one only harvest ever known under the skies that had not been sown" (p. 267).

Theirs is the imagery of Revelation, and of Blake and Carlyle, twisted to absolve those who had crushed the Madame Defarges of France into monstrous forms of responsibility for the monsters they had created. Monseigneur and his class would recast Hosea's prophecy into the essentially atheistic dictum, "they have sown no wind, yet they have reaped the whirlwind." History is reduced to happenstance, indifferent to justice, human or divine.

The narrator's explanation of events, which is to say the philosophy of history underlying the novel, rejects Monseigneur's view. Wending his way among the emigrés and their British supporters at Tellson's Bank, Charles pauses to rebut Stryver's loud abuse of the people in revolution, "when the thing that was to be went on to shape itself out" (p. 267). The thing that was to be is Gabelle's letter, addressed "To Monsieur heretofore the Marquis St. Evrémonde" (p. 268), and laid by the House on Mr. Lorry's desk coincidentally in its addressee's presence. The thing that was to be should also be understood, however, as the course of the Revolution, not only as it threatens Charles and his loved ones but as it generates from the decadence and cruelty of the old regime, vents in the emergent rage of the people, and finally dissipates through the madness by which it consumes itself. In Carlyle's words, "The Revolution then is verily devouring its own children? All Anarchy, by the nature of it, is not only destructive but *self*-destructive" (*Fr. Rev.* 3: 254). When Defarge pleads for the lives of Charles's family, proposing that "one must stop somewhere . . . the question is still where?" his wife retorts, "At extermination"; and her satellites concur: " 'Magnificent!' croaked Jacques Three. The Vengeance, also, highly approved" (p. 369).

History unfolds causally, by a process of reaping what one sows—divine justice indeed. In the fallen world—Paris and London outside the circle of Lucie's golden thread—evil is contagious; it turns victim into perpetrator, morally corrupting, physically threatening all alike. The sycophants who attend Monseigneur at his gala (it is 1780) are characterized as disfigured by "the leprosy of unreality" (p. 137). The image repeats a crucial image in Blake's Prophetic Books, which Dickens had (presumably) never heard of, much less read. And insofar as his diagnosis of the ills that brought the old regime down stems from a source other than his own mother wit, it is, as his rhetoric suggests, attributable to Carlyle:

> Projectors who had discovered every kind of remedy for the little evils with which the State was touched, except the remedy of setting to work in earnest to root out a single sin, poured their distracting babble into any ears they could lay hold of, at the reception of Monseigneur. Unbelieving Philosophers who were remodelling the world with words, and making card-towers of Babel to scale the skies with, talked with Unbelieving Chemists who had an eye on the transmutation of metals, at this wonderful gathering accumulated by Monseigneur. Exquisite gentlemen of the finest breeding, which was at the remarkable time—and has been since—to be known by its fruits of indifference to every natural subject of human interest, were in the most exemplary state of exhaustion, at the hotel of Monseigneur. (pp. 136–37)

Monseigneur's hotel is Dickens' Armida Palace, "where," as Carlyle describes the edifice Louis XV had constructed around Dubarry, "the inmates live enchanted lives; lapped in soft music of adulation; waited on by the splendours of the world;—which nevertheless hangs wondrously as by a single hair" (*Fr. Rev.* 1: 4). Monseigneur's entourage, Louis's court, seek to maintain their own *aevum*: if not to arrest what Dickens himself calls "the current of time" (p. 303), then to insulate themselves from its erosive effects.

In Forests of Night

This effort by the privileged few in both London and Paris consciously or unconsciously to remove themselves from history is reflected in the generic tension built into the novel between history as a narrative mode and Gothic romance. Contemplating the tumbrels as they rumble toward the Place de la Révolution, the narrator enjoins Time to "Change these back again to what they were, thou powerful enchanter . . . and they shall be seen to be the carriages of absolute monarchs, the equipages of feudal nobles, the toilettes of flaring Jezebels, the churches that are not thy father's house but dens of thieves, the huts of millions of starving peasants!" Yet he acknowledges the impossibility of such a reversal: "The great magician who majestically works out the appointed order of the Creator, never reverses his transformations" (p. 399).

Magic, enchantment—recurrent presences in the world of *A Tale of Two Cities*—are perceived to be no more than metaphor, or to be the workings of nature. Neither nature nor history yields to miracle—though the citizens of Saint-Antoine, who act as if each day were the first day, and Lucie and her circle, who act as if no day could alter the lives they had been living the previous day, seem at moments subject to miracle. They seem simultaneously inside and outside history.

Privileged because he had been a victim of the Bastille, Dr. Manette, in quest of the key to Charles's release, can minister to the prisoners in La Force and confront revolutionary authority untouched by the Terror. "I have," he assures Mr. Lorry, "a charmed life in this city" (p. 290). His charmed life, manifesting to Dickens "strength and power" nourished by suffering, is the exact contrary of his burial alive as the nameless cringing shoemaker of One Hundred and Five North Tower, "stopped," as Mr. Lorry imagines him, "like a clock, for so many years" (p. 300).

Black magic, as well as white, characterizes human conduct. If Mr. Lorry's intervention and Lucie's charms restore Dr. Manette to his time-conscious capable self, "set [him] going again with an energy . . . dormant during the cessation of its usefulness" (p. 300), Monseigneur's diabolism reduces the Madame Defarges of France to bestiality and Monseigneur himself to permanent burial or penniless exile: "Like the fabled rustic who raised the Devil with infinite pains, and was so terrified at the sight of him that he could ask the Enemy no question, but immediately fled; so, Monseigneur, after boldly reading the Lord's Prayer backwards for a great number of years, and performing many other potent spells for compelling the Evil One, no sooner beheld him in his terrors than he took to his noble heels" (pp. 263–64).

The ultimate transformation of self and time—worked in Dickens' two cities by magic black and white—is death. To stop like a clock is, finally, to end like the Marquis St. Evrémonde, whom Gaspard's knife turns into the stone face the

Marquis has always resembled. As R. D. Laing remarks on divided selves, in an insight Dickens brilliantly anticipates, petrifaction is a way of killing.[23]

Divisions within selves are, as most critics of the novel have observed, a major focus of *A Tale of Two Cities*. Gaspard's reduction of the Marquis to stone makes physically apparent the emotional and moral petrifaction into which the Marquis had long since frozen.

The Marquis is in this sense descended from a long line of Gothic villains whose villainy turns them to stone. In death, he, like Poe's Roderick Usher, essentially becomes his house, itself to be destroyed by revolutionary vengeance. Arac specifies this confluence of owner and edifice as a device Dickens appropriated from Gothic romance, the popularity of which mushroomed during the years after the French Revolution. He adds that the imagery of deteriorating houses, allegorized as houses of state, invades historical and political writing of the period. Scott, he notes, compares the French monarchy "to an ancient building . . . decayed by the wasting injuries of time"; and Hazlitt deploys the same analogy: "When a government, like an old-fashioned building, has become crazy and rotten, stops the way of improvement, and only serves to collect diseases and corruption, the community . . . 'pull down the house, they abate the nuisance.' "[24]

In *A Tale of Two Cities*, the community pulling down the house, abating the nuisance, takes the form of the incendiaries as they torch the Marquis' chateau, the citizens of Saint-Antoine as they raze the Bastille, and the mob as they slaughter prisoners outside La Force. The mob—ironically, like its mortal enemy, the Marquis—resembles a kind of collective Gothic villain. Masao Miyoshi observes of the Gothic villain (in a comment as applicable to Carlyle's revolutionaries as to Dickens') that he acts out the crippling split within him through violent shifts between devilry and humanity.[25] When Dr. Manette is called to aid an unfortunate suspect, freed, then stabbed by mistake in the street, the doctor finds his patient "in the arms of a company of Samaritans [a moment ago, assassins], who were seated on the bodies of their victims," and who, "with an inconsistency as monstrous as anything in this awful nightmare . . . helped the healer, and tended the wounded man with the gentlest solicitude" (p. 300). When Charles is named at his first ordeal by revolutionary tribunal as the husband of Dr. Manette's daughter, he is viewed through tears that roll down "several ferocious countenances which had been glaring at the prisoner a moment before, as if with impatience to pluck him out into the streets and kill him" (p. 312). When he is acquitted, he is carried home a hero.

The kind of split afflicting the mob and the Marquis is often dramatized by polar characters in conflict rather than characters in conflict with themselves, or even by the opposition between realms of idyllic sunlight and demonic darkness that comprise what Northrop Frye calls the mental landscape of romance[26]—manifest in *A Tale of Two Cities* in Lucie and Madame Defarge, Soho and Saint-Antoine. And Dickens, as if to underscore the acuteness of the threat this split poses for his hero, has Charles act out both versions of its pathology. Paired in the structure of the novel with Carton, he also bears two names: Evrémonde to evoke his past in France, Darnay to embrace his present and future in England. Lawrence Frank, in a provocative essay on *A Tale of Two Cities*, argues that, while Charles disavows his birthright to protest the sins of the past, he commits a comparable (if, for Dickens, venial) sin by denying the responsibilities his birthright imposes on him.[27] Gabelle's plea to Monsieur "heretofore the Marquis" entraps Charles between conflicting obligations: to family and to loyal retainer.

He is caught in what Frank terms an "impasse," thereby epitomizing the dilemma of fallen man. It is in this sense that Evrémonde is "Everyman," required to suffer for the sins of his fathers, and that *A Tale of Two Cities* incorporates an allegory of universal history.[28] Charles's ultimate escape from his impasse through Carton's Christlike sacrifice (vexing to many critics of the novel) is symbolically apt. Though the punishment demanded by revolutionary justice is, as Stoehr reads it, consistent with the dream logic the narrator repeatedly proposes to account for Paris in upheaval, it is disproportionate to Charles's real guilt. Thus his double, Carton, must expiate his guilt.[29]

Frye cites Carton's martyrdom as an example of the archetypal pattern in romance whereby one character sacrifices himself to prolong or renew the life of another.[30] The plot of *A Tale of Two Cities* consists of a series of such fateful, if mainly less drastic, interventions in the lives of its characters, of which Carton's intervention in the life of Charles is the climactic and most dramatic. Each of these interventions alters or integrates divided selves. The series starts with the intervention of Lucie and Mr. Lorry to disinter Dr. Manette, entombed within the broken shoemaker of One Hundred and Five North Tower. His rescue leads to Lucie's intervention in the life of Mr. Lorry.

"I am," Mr. Lorry says, broaching what he represents to Lucie as the story of a customer, "a man of business. I have a business charge to acquit myself of. In your reception of it, don't heed me any more than if I was a speaking machine—truly, I am not much else" (p. 54). John Kucich argues that as a man of business, Mr. Lorry embodies an institutional violence which imposes social order in England as revolutionary violence erodes social order in France: that the name "Lorry" constitutes a pun, evoking the tumbrels on their way to the guillotine, and that Mr. Lorry himself enacts the consequences of repression.[31] Mr. Lorry acknowledges to Lucie his dehumanization under Tellson's discipline, recognizing that it is intended to drain him of all potential for human feeling: "These are mere business relations, miss; there is no friendship in them, no particular interest, nothing like sentiment. I have passed from one to another, in the course of my business day; in short, I have no feelings; I am a mere machine" (p. 54).

Lucie cannot be a mere machine; she cannot, despite Mr. Lorry's plea ("Pray control your agitation—a matter of business" [p. 55]), repress her feelings. Mr. Lorry himself concedes as much when he divides the labor involved in unburying Dr. Manette. He and Lucie go to Defarge's wine shop, he explains, "I, to identify him if I can: you, to restore him to life, love, duty, rest, comfort" (p. 57).

This division of labor is also reflected in the tension built into the narrative between history and romance. Summing up his disclosures to Lucie—"The best and the worst are known to you, now" (p. 57)—Mr. Lorry echoes the narrator, who introduces events by summing up the best and the worst in 1775, the year their adventure begins. The detachment with which the narrator views history, or Mr. Lorry his quest for Dr. Manette, is impossible not only for Lucie and her father but (as he gradually learns) for Mr. Lorry to maintain. Their dilemma determines the narrative focus, which subordinates public concerns to the personal trials of hero, heroine, and supporting cast. As Dr. Manette traverses Paris, exhausting stratagem after stratagem to free Charles, the Revolution swirls around him:

> The new era began: the king was tried, doomed and beheaded; the Republic of Liberty, Equality, Fraternity, or Death, declared for victory or death against the world in arms; the black flag waved night and day from the great towers of Notre Dame; three hundred thousand men, summoned to rise against the tyrants of the earth, rose from all the varying soils of France, as if the dragon's teeth had been sown broadcast, and had yielded fruit equally on hill and plain, on rock, in gravel, and alluvial mud, under the bright sky of the South and under the clouds of the North in fell and forest, in the vineyards and the olive-grounds and among the cropped grass and the stubble of the corn, along the fruitful banks of broad rivers, and in the sand of the sea-shore. (p. 301)

These monumental events, which occupy Carlyle for hundreds of pages, are to Dickens little more than notations on a calendrical grid, against which his domestic romance is played out.

Georg Lukács attacks *A Tale of Two Cities* on just this ground. To him, it fails as a historical novel because it reduces history to a backdrop for the drama of Lucie, Charles, Carton, and Dr. Manette.[32] But that is precisely Dickens' intention. "What private solicitude," the narrator asks of Dr. Manette's frustrated efforts on behalf of Charles, "could rear itself against the deluge of the Year One of Liberty" (p. 301)?

To Dickens, history and political conflict offer no cure for human ills. The one historical document established as germane, even crucial, to the dynamics of his plot—Dr. Manette's prison narrative—is used to justify a crime as great as the crime it condemns. Only the power to connect, which E. M. Forster was later to claim for "personal relations," eases pain in the cities of Dickens' *Tale*. Lucie's love restores Dr. Manette; the friendship of Lucie and her father restores Mr. Lorry. The man who professes himself a machine, without feeling, participates in what appears to him almost "a horrible crime" (p. 235)—the dismembering and burning of the shoemaker's workbench—to protect Dr. Manette from his other self. In the effort to save Charles from the vengeance of Madame Defarge, Mr. Lorry allows his feelings for Dr. Manette's family to eclipse his function as an agent of Tellson's.

And he abets Lucie in rescuing Carton from his own despair. That Mr. Lorry and Carton are bachelors, and Carton an orphan, is in the fitness of things as much as Madame Defarge's death at the hands of Miss Pross. Neither comes to this crisis with any personal relations to guide or support him. Both have, like Charles, if for different reasons, repudiated their pasts, even their futures, to serve as mere tools of profit in a restrictive present: Carton as a tool of Stryver, Mr. Lorry of Tellson's. When Carton recognizes in Mr. Lorry a surrogate father—"I could not see my father weep, and sit by careless. And I could not respect your sorrow more, if you were my father" (p. 338)— and Mr. Lorry assents to this unaccustomed role—"He gave him his hand, and Carton gently pressed it" (p. 338)—they form a human connection from which they had been cut off.

In a perceptive article on *A Tale of Two Cities*, Albert Hutter suggests (though he does not explicitly argue) that the search of Mr. Lorry and Carton for emotional ties, which they finally realize as surrogate father and son, makes part of a narrative structure paralleling the strains in the family and in society produced by Victorian ambivalence toward authority. Authority is, in characteristic Freudian fashion, embodied in fathers.[33] For Carton to embrace Mr. Lorry as a father is to acknowledge the moral order of society and the ultimate source of order, God. It is after leaving Mr. Lorry that Carton recalls the words spoken at his own father's grave, "I am the resurrection and the life" (p. 342), and irrevocably decides on his sacrificial course.

His meeting with Mr. Lorry also leads to the reform of Carton's parody double as resurrection man, and the bearer of divine initials, Jerry Cruncher. Hutter treats the comic subplot that centers on Cruncher as part of a primal scene fantasy, pitting fathers against sons in a tension that pervades the novel. As Solomon Barsad and Roger Cly, agents of repressive paternalistic regimes in both England and France, spy on Charles, on Saint-Antoine, and eventually on the prisoners of the Revolution, young Jerry spies on his resurrection man father plying his nocturnal trade.[34]

The resurrection man plying his trade is himself—at variance with the symbolism the Freudian scheme assigns to fathers—rebelling against authority. As Mr. Lorry admonishes Jerry, on witnessing his exposure of the hoax by which Cly was recalled to life, "You have used the respectable and great house of Tellson's as a blind, and . . . you have had an unlawful occupation of an infamous description" (p. 335–36). Jerry has subverted the ethical and legal codes Mr. Lorry represents.

The subversion of those ethical and legal codes, however, brings about a larger good, for through it Jerry comes by the information Carton uses to force Barsad to cooperate in Charles's rescue. The novel's imitators of Christ (one comic, one tragic) join to become resurrection men indeed. And in joining to resurrect Charles, they redeem themselves. As Carton gives his life to save the life of his rival for Lucie, and his spiritual support to cheer the little seamstress in their last moments, Jerry vows that if the Darnays and Dr. Manette escape, he will give up his unlawful occupation and stop interfering with his wife's prayers.

Hutter characterizes Jerry's attack on Mrs. Cruncher—when frustrated because he has found Cly's grave empty, he knocks "her head against the head board of the bed" (p. 192) while young Jerry watches—as a parody of rape.[35] Whether we are altogether persuaded by Hutter's Freudian reading (I find his evidence a bit thin), Jerry's private violence against his wife mirrors Saint-Antoine's public violence against king and country. Living hand-to-mouth in London provokes Jerry to beat Mrs. Cruncher for (he supposes) thwarting his honest trade by prayer; living hand-to-mouth in Paris provokes the citizens of Saint-Antoine to massacre all they suppose guilty of withholding bread from them. Though Madame Defarge may kill to avenge her sister and brother, and Gaspard to avenge his child, the visceral motive of the revolutionaries collectively is imaged (as Dickens would have learned from Carlyle, if from no other source) in the meager provisions the mender of roads brings to his daily work or in the pinched faces around the fountains in Saint-Antoine and the village below the Evrémondes' chateau.

Hunger reduces men to beasts. In English society, even if displaced temporarily to the wrong side of the Channel, the intervention of his benevolent betters (chiefly Mr. Lorry) teaches Jerry a moral lesson and restores his humanity; in French society the intervention of their malevolent betters reinforces the Sansculottes' savage instincts. The Marquis in his chambers is described as moving "like a refined tiger" (p. 156), counterpart of the tiger the narrator repeatedly perceives in Madame Defarge. Dickens' France resembles Blake's Europe: forests of night (or eternal death) inhabited by lions and tigers, consumed and consuming, devouring and being devoured.

Since the novel's recurrent metaphor for France in upheaval is nature red in tooth and claw, the appearance of *A Tale of Two Cities* in 1859, the same year as *The Origin of Species*, seems an almost magical coincidence. Crushing humanity out of shape is tantamount to a kind of devolution. The narrator sees "in the hunted air of the people ... some wild-beast thought of the possibility of turning at bay" (pp. 61–62). Gaspard, using the lees of Defarge's spilled wine to scrawl BLOOD on a wall in Saint-Antoine, is described by the narrator as "a joker of an extremely, not to say wolfishly practical character" (p. 63). Gaspard, lamenting his child, howls "over it like a wild animal" (p. 141). The people are, to the Marquis, rats, dogs, or pigs. Charles rearrested is a type of Actaeon (a figure Dickens may have drawn from Carlyle), his tribunal "a jury of dogs empanelled to try the deer" (p. 345). His wife and daughter are, to Madame Defarge, no more than "her natural enemies and her prey" (p. 391).

Carton playing jackal to Stryver's lion suggests that England, despite the humanizing influences exerted on his life and on Jerry's, is potentially no less a forest of night than France. The fair city Carton envisions when he leaves Stryver's flat emerges from the "wilderness before him" (p. 121). The interest that the crowd at the Old Bailey takes in Charles's trial for treason is "Ogreish" (p. 93); the mob that attacks Cly's funeral coach is "a monster much to be dreaded" (p. 186).

Among critics of the novel, the standard response to such details is to read in them Dickens' warning to his countrymen that revolutionary unrest simmers just beneath the surface on their side of the Channel too. As near to its composition as April, 10, 1855, he wrote to A. H. Layard:

> I believe the discontent to be so much worse for smouldering instead of blazing openly, that it is extremely like the general mind of France before the breaking out of the first Revolution, and is in danger of being turned by any one of a thousand accidents—a bad harvest—the last strain of too much aristocratic insolence or incapacity—a defeat abroad—a mere chance at home—into such a Devil of a conflagration as never been beheld since.[36]

Though the explosive potential he detected in English society was preoccupying Dickens even as he embarked on *A Tale of Two Cities*, to treat the novel as mainly his letter to Layard encoded in a fiction oversimplifies his design and understates his achievement. John C. Greene points out that Social Darwinism actually preceded Darwin.[37] And three years after *A Tale of Two Cities* and *The Origin of Species* appeared, Herbert Spencer in *First Principles* proposed directly that Darwinian evolution, natural selection, operated in society as well as in nature:

> The authority of the strongest makes itself felt among a body of savages, as in a herd of animals, or a posse of school-boys. At first, however, it is indefinite, uncertain; it is shared by others of scarcely inferior power, and is unaccompanied by any difference in occupation or style of living; the first ruler kills his own game, makes his own weapons, builds his own hut, and economically considered, does not differ from others of his tribe. Gradually, as the tribe progresses, the contrast between the governing and the governed grows more decided. Supreme power grows hereditary in one family; the head of that family, ceasing to provide for his own wants, is served by others; and he begins to assume the sole office of ruling.[38]

While everything about Dickens indicates that he would have loathed Social Darwinism, at least as a guide for public policy, his persistence in comparing Paris and London, both terrorized by mobs, to nature suggests an organic view of society—of history—close to Spencer's own. In tracing "social dissolution" to circumstances that develop "when social evolution has ended and decay has begun," Spencer might well have been describing Dickens' France.[39]

Spencer assumes evolution and devolution to be parts of an inevitable dialectical process that constitutes history. Though *A Tale of Two Cities* ends happily—Carton is apotheosized, the fugitives escape, a brighter future is predicted for France—the novel's resolution implies a similar process. The price of the Darnays' survival is the sacrifice of Carton and the relapse of Dr. Manette. As Kucich observes, moreover, the novel unfolds by a series of such transactions. The rescue of some characters requires the abandonment of others: Lucie as a child is saved by Mr. Lorry at the cost of leaving her father behind; Charles is saved at the cost of having to leave Carton (and, temporarily, Jerry and Miss Pross) behind.

Kucich traces this compensation of gain with loss to the repressive powers at work mainly in France.[40] In a Marxist world, repression would yield to revolution. In Dickens' world, revolution proves to be repression by another name. Madame Defarge is no less a tyrant than Monseigneur. For Dickens' world is the world of Adam's Fall. The sole way out of the impasse in which it immures its inhabitants is the way out of history altogether. That Charles and his family, accompanied by Mr. Lorry, leave history to return to their *aevum* in Soho is hinted at by their abrupt disappearance from the narrative. Yet they cannot escape history: they bear it with them in their own mortality, in their memory of Carton, and in the pathetic shape into which Dr. Manette has been crushed. The sole escape from history is, finally, Carton's escape: to a world—perhaps fantasy, perhaps reality—where there is no time, and no trouble.

SIX

Proving Nothing:
Hardy's
The Dynasts

The Burthen of the Mystery

"I left off on a note of hope," Hardy wrote to his friend Edward Clodd in 1908. "It was just as well that the Pities should have the last word, since . . . *The Dynasts* proves nothing."[1] Thus, with a single pronouncement, he anticipates and would apparently have dismissed almost eighty years of critical debate concerning the metaphysical significance of the Overworld in his drama of the Napoleonic Wars.

That Hardy meant what he told Clodd is underscored by his explanation of the Spirits in the preface he wrote to Part First in 1904: "They are intended to be taken by the reader for what they may be worth as contrivances of the fancy merely. Their doctrines are but tentative, and are advanced with little eye to a systematized philosophy warranted to lift 'the burthen of the mystery' of this unintelligible world" (p. xxiv). The Spirits reflect various of the nineteenth century's speculations on the dynamics of history, both natural and human. They embody, as several critics have pointed out, mental constructs. As Hardy put it in a letter to Clodd, also written in 1904, "They are not supposed to be more than the best human intelligence of their time in a sort of quintessential form."[2]

Though it tells a story that is satisfyingly complete and (in large part) historically true, *The Dynasts* remains metaphysically open, "tentative." The Immanent Will may, as the Chorus of the Pities predicts, develop consciousness and, with it,

Christlike sympathy for human pain. But the Spirit of the Years only momentarily, and never more than half, believes it: *"You almost charm my long philosophy / Out of my strongbuilt thought"* (After Scene, p. 524); and the response to Years's skepticism, antiphonally chanted by the Semichoruses of the Pities, is largely interrogative and problematic:

SEMICHORUS I OF THE PITIES

Nay;—shall not Its blindness break?
Yea, must not Its heart awake,
    Promptly tending
    To Its mending
In a genial germing purpose, and for loving-kindness' sake?

SEMICHORUS II

Should It never
Curb or cure
Ought whatever
Those endure
Whom It quickens, let them darkle to extinction swift and
    sure.

(After Scene, p. 525)

Against such uncertainty and the ten years of blood that supply the action of *The Dynasts*, the promise in which the semichoruses join to close the curtain (*"That the rages / Of the ages / Shall be cancelled, and deliverance offered from the darts that were, / Consciousness the Will informing, till it fashion all things fair"*—After Scene, p. 525) seems at best the precarious hope Hardy labeled it in his letter to Clodd.

By leaving unanswered the metaphysical question the Spirits debate, Hardy contrasts the historical circumstances of the Napoleonic Wars, discoverable from myriad accounts (many of which he had read), with the mystery of this unintelligible world. The gap between data accessible to human intellect and the inaccessible forces that cause and order—between what and why—is the territory Hardy explores in *The Dynasts*. To leave unanswered the metaphysical question the Spirits debate also, as he well knew, was the stance taken by most of his scientific contemporaries toward first and last things. Clodd had written in his *Pioneers of Evolution from Thales to Huxley* (1897) that science "may borrow the Apostle's words, 'Behold! I show you a mystery,' and give to them a profounder meaning as she confesses that the origin and ultimate destiny of matter and motion; the causes which determine the behavior of atoms ... ; the conversion of the inorganic into the organic by the green plant, and the relation between nerve-changes and consciousness; are all impenetrable mysteries."[3] An impenetrable mystery is how Years describes the Immanent Will, dismissing the possibility that It may act from comprehensible, humanly rational, motives:

> It works unconsciously, as heretofore,
> Eternal artistries in Circumstance,
> Whose patterns, wrought by rapt aesthetic rote,
> Seem in themselves Its single listless aim,
> And not their consequence.

(Fore Scene, p. 1)

Years's description of the Immanent Will suggests Hardy's perception that in mechanistic cosmology, while accident recedes ever farther toward the germinal core of the universe, its structure cannot finally be rationalized.

For Years is the consummate mechanist. To him, the universe operates by "Clock-like laws" (Fore Scene, p. 1): though the first cause appears to be unfathomable, the increasingly complected chains of effects and secondary causes, triggered by the first cause, unwind of necessity.[4] Hardy may have drawn this deterministic understanding of nature and of man in part from Tolstoy, who posits a world clocklike in its operation to explain the Battle of Austerlitz. But he has Years propose a view of the dynamics of history that, in Great Britain, was perhaps most strongly advocated by Henry Thomas Buckle. Buckle defines as the aim of his History of Civilization in England (1856–1861) the achievement of "something equivalent, or at all events analogous, to what has been effected by other inquiries for the different branches of natural science." He argues that in human affairs, as in nature, phenomena apparently arbitrary actually obey "certain fixed and universal laws."[5]

Convinced that only religion, with its faith in Providence, keeps historians from the discovery of fixed and universal laws for history, Buckle, however, oversimplifies the discoveries of science. Hardy, indeed, attacks Buckle's historical method, asking in his journal, "Is not the present quasi-scientific system of writing history mere charlatanism?"[6] In *The Dynasts* he depicts this quasi-scientific system, causal logic, as a strategy to free political conduct from moral constraint. Urging the scheme to assassinate Napoleon on a scandalized Fox, Guillet de la Gevrillière reasons that "The sovereign remedy for an ill effect / Is the extinction of its evil cause" (II, i, i, 145).[7]

And in *First Principles*, which Hardy was enthusiastically to recommend to Lena Milman, Spencer turns Buckle upside down. Conceding "the power . . . the universe manifests . . . [to be] utterly inscrutable," he insists on the need not to divorce but to reconcile science and religion.[8] The inscrutability, the imagelessness, of that power poses for the scientist, or scientific historian, the problem of finding a language, an imagery, the reader can grasp, without at the same time anthropomorphizing, if not deifying, the process being described. Darwin himself confronts this problem:

> It has been said that I speak of natural selection as an active power or Deity; but who objects to an author speaking of the attraction of gravity as ruling the movements of the planets? Everyone knows what is meant or is implied by such metaphorical expressions; and they are almost necessary for brevity. So . . . it is difficult to avoid personifying the word Nature; but I mean by Nature, only the aggregate action and product of many natural laws, and by laws the sequence of events as ascertained by us.[9]

Darwin's difficulty in finding a metaphor by which to define natural selection, without misrepresenting it as an active power or deity, is similar to Years's difficulty in persuading the Spirit of the Pities that it makes no sense to blame the Immanent Will for human suffering. Years characterizes the Will as a mind that determines without foreknowledge, or forethought, and proffers label after label to describe It (as if testing them on his own ear): "*processive, / Rapt, superconscious; a Clairvoyancy / that knows not what It knows, yet works therewith.*" But he must finally surrender to the inadequacy of language itself: "*The cognizance ye mourn, Life's doom to feel, / If I report it meetly, came unmeant, / . . . luckless, tragic*

Chance, / In your more human tongue" (I, v, v, 99–100). His definition of the Will's aims—as patterns wrought by unfolding circumstance, rather than their consequences—suggests Darwin's definitions: of nature as the aggregate action and product of natural laws, of law as the sequence of events we ascertain. These constructs approach modern ideas of symbolism, which demand that an image mean primarily itself, and that no rigorously delimited relation can exist between an abstract tenor and concrete vehicle. The workings of the Will are, to Years, "artistries."

Proclaiming the Will an artist, his workings (history and, I suppose, natural history) artistries, is analogous to perceiving in the rhetorical design of *The Origin of Species* a form of artifice, a case for which Hayden White and, before him, Stanley Edgar Hyman have both made. Darwin's key phrase, "the struggle for existence," should, Hyman argues, be read as a metaphor that heightens a progress of infinitesimal gradations. *The Origin of Species* is a "dramatic poem of a special sort," positing a world of adversaries locked in endless war.[10] Darwin's nature suggests Hardy's Europe. The French fugitives who drown beneath the last usable bridge across the Berezina are (in a striking echo of Scott) *"the weak pushed over by the strong"* (III, i, xi, 356). *The Dynasts*, repeatedly reducing men and armies to mackerel, minnows, mollusks, caterpillars, or snakes, analogizes Europe to Darwinian nature.

Hardy's encapsulation of history reflects the kinship between romantic epistemology and scientific method. Both are rooted in experience, differing only in that scientists select and analyze rather than try to digest experience whole.[11] That this practice of scientists tends to isolate phenomena, and is therefore reductive in effect, is partly what White means when he nominates metonymy ("the substitution of the name of a part of a thing for the name of the whole") as the "favored trope of *modern* scientific discourse."[12] He is translating into technical terms a view of scientific discourse that roughly approaches Darwin's: metaphor, though an essential shorthand, must be drained of any implication that a spiritual presence underlies natural events. Darwin does not necessarily deny God's existence; he doubts (as Hardy's conception of an Immanent Will expresses his doubts) that God operates by humanlike cerebrations. "Have we any right to assume," Darwin asks, "that the Creator works by intellectual powers like those of man?" And he excludes from the purview of science the search for first causes: "I may here premise that I have nothing to do with the origin of the mental powers, any more than I have with that of life itself."[13]

Closing Up Time and Traversing Space

Darwin's confinement of science to the investigation of observable phenomena illustrates White's claim that scientific discourse, because it is reductive (and therefore metonymic), rejects teleology for contiguity.[14] Science, as Robert Langbaum puts it, selects and rationalizes.[15] Treating history not scientifically but as the stuff of art, Hardy selects and rationalizes too. In the closing speech of the Fore Scene to *The Dynasts* (p. 7), the General Chorus of Intelligences announces that

> *We'll close up Time, as a bird its van,*
> *We'll traverse Space, as spirits can,*
> *Link pulses severed by leagues and years,*
> *Bring cradles into touch with biers;*
> *So that the far-off Consequence appears*
> *Prompt at the heel of foregone Cause.*

As Susan Dean points out, the Intelligences prefigure the technique of *The Dynasts* overall.[16] Time and space, to the Spirits that journey between Salamanca and Borodino, and to the imagination accompanying them, are *"A fancy"* (III, i, iv, 339). To close up time and traverse space is to "vision onward" (I, i, vi, 32), to engage in what Blake would call mental traveling. When the Intelligences acknowledge that "The PRIME, that willed ere wareness was, . . . We may but muse on" (p. 7), their pun on "muse" implies Hardy's rejection of positivism's stance that the application of scientific method to social and political life will yield insight into the dynamics of human history comparable to the insight achieved by science into the dynamics of natural history. Susceptible neither to generalization nor to quantification, human experience is to be explored at best tentatively, by the methods of the Muses.

For history is story. The Spirit Sinister hopes Napoleon will invade England because *"Peace is poor reading"* (I, ii, v, 54). The Chorus of the Years turns the attention of the Spirit audience from Nelson's triumph at Trafalgar to Napoleon's counterstroke at Austerlitz, remarking how *"fiercely the predestined plot proceeds"* (I, v, vii, 108). Napoleon at least dimly senses himself a protagonist in someone else's tale. And death, to which he fears he will go before he produces an heir, is defined by the Chorus of the Pities as a locale *"where History pens no page"* (III, i, xi, 357).

To pen pages of history is to "emplot" it. The term is White's. He is probing not the strategies of poets who use history for their own designs but the strategies of the historians themselves; and he argues that choosing where to begin and end amounts to imposing artistic form on events.[17] The symmetry of the beginning and ending of Hardy's "plot" in *The Dynasts* suggests as much. The action that eventuates in the epic encounter between Napoleon and Wellington essentially starts with Parliament's debate, at midnight, of Pitt's "Provision for England's Proper Guard" and stops with Napoleon alone, at midnight, in the Wood of Bossu. I say "starts" and "stops" because the bounds Hardy sets for his drama are, from a historical perspective, arbitrary. He might have opened with 1789, as Scott does his *Life of Napoleon*; or with his hero's background and birth as Hazlitt does his *Life*; or with the First Coalition against revolutionary France; or with Bonaparte's emergence as a soldier of the Revolution. He might have closed with Napoleon's exile to St. Helena or with his death. Instead Hardy focuses on the thirteen years between the resumption of war after the Peace of Amiens and Waterloo, a period that unfolds as a coherent whole in *The Dynasts* because Hardy, not history, renders it so.

He renders it so by selecting and juxtaposing events to make them reverberate beyond themselves. He closes up time and traverses space. Part Second of *The Dynasts* moves from Napoleon's triumphs at Jena and Auerstadt to his entry into Berlin, omitting the decisive battles of Eylau and Friedland. For Hardy, whose intent is not to chronicle Napoleon's rise but to capture the forces that made it happen, Eylau and Friedland are redundant. I do not mean that they are ignored, as if they never happened (though Hardy is capable of that too). The Semichoruses of the Pities recall both battles in their account (II, i, vii, 168) of how the "great Confederacy" of the allies breaks down.

These Semichoruses, and most of the other exchanges that take place in the Overworld, provide the kind of commentary on the action that conventional historians provide when they delineate cause and effect to explain how events cohere and form a unified whole. White observes that historical narratives typically have two levels: a story and a conceptual analysis, which emerges from the story and establishes the type of story it is.[18] The story in *The Dynasts* is the drama of the Napoleonic Wars; its analysis consists of the Spirits' responses to the drama. This second level, White argues, is where the author's mythic consciousness chiefly asserts itself. White builds on Northrop Frye, who distinguishes between the chronicler, for whom temporal sequence makes a sufficient narrative rationale, and the historian, for whom narrative rationale entails a structure shaped to a "conceptual framework" derived not from life but from literary tradition, and so ultimately from myth.[19] What "myth" implies for Frye is anticipated by the Spirit of the Pities and the Spirit Ironic as they debate (Fore Scene, p. 4) whether they are witnessing tragedy or comedy. Their difference over which type of play history is reflects Hardy's recognition that, broadly enough conceived, the historian's scheme may conform to a literary model.

White develops his comparable recognition that historians construct their stories according to fictive modes into a rigorous system for defining the genres of historical narratives. This recognition leads him to propose that, because it melds a manifest or literal surface (the sequence of events or "plot") with a latent subsurface (the metahistorical or mythic rationale that underlies plot), historical narrative resembles dream.[20] Hardy too detects this resemblance. Semichorus II of the Pities images the Prusso-Russian alliance, after the battles of Eylau and Friedland, *"like the diorama of a dream"* (II, i, vii, 168). The debacle at Leipzig is, to Napoleon, a nightmare lived: "Am I awake, / Or is this all a dream?—Ah, no. Too real!" (III, iii, iv, 388). Waterloo is, to the soldiers on both sides, another nightmare lived: "By the lurid light the faces of every row, square, group, and column of men, French and English, wear the expression of those of people in a dream" (III, vii, vii, 505).

The contorted phrasing of Hardy's stage direction invites us to see these masses both as congeries of individuals wrestling with horrors, and as figures in the dream of the Immanent Will. After the curtain has fallen on Waterloo, Semichorus I of Ironic Spirits sums up history as the Will's *"Fixed Foresightless dream"* (III, vii, 517); and the Spirit Ironic himself specifies the dilemma inherent in the Semichorus's skepticism: *"Who knows if all the Spectacle be true, / Or an illusion . . ."* (After Scene, p. 524)—or, as Blake might have envisioned it, Enitharmon's nightmare.

But *The Dynasts* is not Blake's *Europe.* Despite the persistence of the analogy between history and dream, Hardy's drama, unlike Blake's poem, holds their realms rigidly separate: history is not a dream, it only resembles one; and then it resembles not the dream itself but the diorama of a dream—images from a dream, sometimes acted, more often mechanically reproduced. Hardy has Years describe Napoleon, when he accepts Mack's surrender at Ulm, as moving *"like a figure in a lantern-slide"* (I, iv, vi, 76). This representation of history as a series of projected images, with Hardy's manipulation of visual perspective in *The Dynasts,* has prompted John Wain to credit him with inventing cinematic technique before there was cinema.[21] And Hardy's word to describe the fate of the Prusso-Russian alliance, "dissolves" (suggesting the merger of one episode into the next), is precisely the term a filmmaker might use.

Like film, or most other drama, and like dream, *The Dynasts* compresses time, renders it as something other than segments measured by clock and calendar. Watching by Nelson's deathbed in the cockpit of the *Victory,* Captain Hardy answers his commander's question, "What are you thinking, that you speak no word?"

> Thoughts all confused, my lord:—their needs on deck,
> Your own sad state, and your unrivalled past;
> Mixed up with flashes of old things afar—
> Old childish things at home, down Wessex way,
> In the snug village under Blackdown Hill
> Where I was born. The tumbling stream, the garden,
> The placid look of the grey dial there,
> Marking unconsciously this bloody hour,

And the red apples on my father's trees,
Just now full ripe.

(I, v, iv, 97)

Thomas Hardy has the captain of the *Victory* adapt to his own
needs Carlyle's geometric trope for the problem of the artist
who would write history: that while action is solid, narrative
is linear. As Captain Hardy waits out the crisis of Nelson's
death dragging its agonized length toward completion, he remains aware of the hours indifferently passing, of the year also
indifferently passing, of his crew, oblivious of time passing,
fighting for their lives.

Time, history, are lived variously by various individuals,
their experiences shaped by the circumstances in which they
are caught. In Carlyle's formulation, events, often simultaneous rather than successive, constitute "an ever-living, ever-
working Chaos of Being" (*CME* 2:88)—a phrase evocative of
the way battle is treated in *The Dynasts*. Hardy admired Carlyle, especially his grasp of the epistemological problem that
faces the writer of history. He found the problem succinctly
put by an essayist in *The Spectator* of May 6, 1882, who (in a
statement Hardy copied into one of his notebooks) praised Carlyle for showing "how small a proportion of our life we can
realize in thought; how small a proportion of our thoughts we
can figure forth in words; how immense is the difference between the pretensions of human speech and the real life for
which it stands."[22]

The gap between speech and life is, as J. Hillis Miller
stresses, a central concern in *The Dynasts*. Miller adopts Hardy's view of the relation between the historian and history, arguing that there can be "innocent" (that is, "objective") readings neither of events nor of their narratives. In their attempts
to discover pattern in circumstance, historians, like literary
critics, unravel and reweave the fabric of events, inevitably dis-
torting them in the process.[23]

The acknowledgment that innocent readings are unattaina-
ble, that their unattainability may for the historian even be a
virtue, comprises one of the strands Hardy singled out in Carlyle's work. Among his several notes on Carlyle, he transcribed
(with some changes) J. A. Froude's celebration of *The French*

Revolution as an "Aeschylean drama composed of facts literally true."[24] To Froude, the greatness of Carlyle's *French Revolution* lay in its elevation of history into myth. Hardy would refashion history similarly. In his preface to *The Dynasts*, he suggests that he understands "Aeschylean drama" partly as a dramatic strategy that makes the conceptualization of history possible. Justifying a structure based on the reader's foreknowledge of events, he pronounces it "interesting, if unnecessary to name an exemplar as old as Aeschylus, whose plays are . . . scenes from stories taken as known, and would be unintelligible without supplementary scenes of the imagination" (p. xxv). By way of reminding the reader of his own capacity to supply those supplementary scenes, Hardy establishes his privileged position at the outset. Having placed in Years's mouth French confidence in the power of Napoleon to cross the Channel against England's raw merchant-fellows and laborers (*"in one summer night / We'll find us there!"*), he has the Recording Angel ask, *"is this prophecy true?" "Occasion will reveal,"* Years answers (I, i, ii, 14–15). For the reader, occasion has revealed. Only the Spirits—"contrasted Choruses" Hardy calls them (p. xxv)—must, like Aeschylus' chorus, discover truth by watching the action unfold.

Hardy professes to treat the Napoleonic Wars as Aeschylus treats what we think of as myth. He would transform "facts literally true" about the struggle between France and the allies into a heroic myth that "re-embodies" England's role in its "true proportion." Hardy's word "re-embody" sums up his method of dramatizing England's role. *The Dynasts* depicts England re-embodied in a succession of heroes—Nelson, Pitt, Moore, Wellington—each of whom, with his subordinates, confronts Napoleon. When Pitt is hailed during the celebration of Trafalgar as "the Savior of England," he replies (in words he is recorded as having actually spoken) "No man has saved England . . . / England has saved herself, by her exertions: / She will, I trust, save Europe by her example!" (I, v, v, 103).[25]

That Hardy takes seriously Pitt's claim for England as the savior of Europe is reflected by the tripartite structure of *The Dynasts*, with Sir John Moore's miraculous triumph and tragic death before Coruña at its center (Part Second of three, act iii of six, scenes ii and iii of five). Moore embodies England's heroic example; and whether history sees in his battle the turning point of Napoleon's career, Hardy makes it the turning point of his drama. As Years comments on Napoleon's plan to go to Astorga, "then turn back," leaving to Marshal Soult "the destinies of Spain," *"More turning may be here than he designs. / In this small, sudden, swift turn backward, he / Suggests one turning from his apogee!"* (II, iii, ii, 215). After Coruña, and despite the French victory at Wagram, *The Dynasts* largely consists of Napoleonic defeats, charting Bonaparte's decline from the Peninsula, through Moscow and Leipzig, to abdication and exile on Elba, finally to Waterloo.

In making Coruña the pivotal episode in this sequence, Hardy enters the debate (represented on one side by Scott, who indicts Moore for caution and pessimism, on the other by W.F.P. Napier, who lauds him for courage and prudence) over his conduct of the bitter British retreat from Astorga which culminated in the battle.[26] That Hardy devotes full-blown scenes not only to the British stand at Coruña but to Moore's death and burial (drawn from Napier and from Moore's brother, James C. Moore, in his overt apologia for Sir John) attests to the side he takes.[27] For Moore recalls—in effect "re-embodies"—Nelson. As Nelson, dying, tells Captain Hardy, "I'm satisfied. Thank God, I have done my duty!" (I, v, iv, 98), Moore, also dying, tells Sir John Hope, "I hope that England—will be satisfied— / I hope my native land—will do me justice!" (II, iii, iii, 219).

The difference in their last words reflects the difference between Moore's awareness that his tactics will be questioned and Nelson's that his have earned him a place in England's pantheon of heroes. Nelson's satisfaction at having done his duty identifies him as the ultimate realization of his own demand, signalled to the fleet before the battle: "ENGLAND EXPECTS EVERY MAN WILL DO HIS DUTY" (I, v, i, 83).

"The Curious Literariness of Real Life"

That demand and its fulfillment, epitomizing in the public mind Nelson's conduct at Trafalgar, constitute a piece of history absorbed into myth. Like the experience of the trenches (transformed into literature) that Paul Fussell explores in *The*

Great War and Modern Memory, it has become part of our cultural consciousness. Fussell proposes, as a possible subtitle for his book, "An Inquiry into the Curious Literariness of Real Life."[28] *The Dynasts* too suggests the curious literariness of real life. Hardy took Nelson's last utterance—in which he claims, with the inexorable logic of romance, to have fulfilled the ideal he has urged on his men—from William Beatty's eyewitness account of the event.[29]

But Hardy is also capable, in the interests of symbolic truth, of adjusting real life to fit. To the scene of Villeneuve's suicide he adds a note, explaining that he has antedated it "to include it in the act to which it essentially belongs" (I, v, vi, 104), and he alters details reported by Beatty. Beatty recalls coming to Nelson's aid, to be told by the admiral, "Ah, Mr. BEATTY! you can do nothing for me."[30] Hardy, however, has Nelson confide his diagnosis to Dr. Scott, the ship's chaplain: "Doctor, I'm gone. I am a waste o' time to you" (I, v, iii, 88). One might expect that Hardy would scarcely have had Nelson, in his last moments, declare himself a waste of the chaplain's time. But his Nelson feels assured of the immortality he craves. As Villeneuve, about to commit suicide, declares (in what amounts to an epitaph for his adversary), Nelson is "blest and over blest / In [his] outgoing at the noon of strife / When glory clasped [him] round" (I, v, vi, 104).

Clasped round by glory also characterizes Moore's outgoing. Although he has not become a part of British heroic tradition, as Nelson has, *The Dynasts* would make him so. He too stands in the forefront, demanding ("Forty-second, remember Egypt!" [II, iii, iii, 217]) gallantry from his men. He too, perceiving death in his wound, continues to be more concerned for his army than for himself. He too pronounces judgment on his own end: " 'Tis *this* way I have wished to die!" (II, iii, iii, 212). He too has his epitaph spoken by a foreigner:

> His was a spirit baffled but not quelled,
> And in his death there shone a stoicism
> That lent retreat the rays of victory.
>
> (II, iii, v, 224)

That the source of these words is a sympathetic Viennese civilian rather than an opponent in the field measures the extent to which Hardy portrays Moore as, however courageous, nonetheless a decline from Nelson. Though Hardy knew from Napier that Soult had erected a monument to Moore, he withheld this detail from *The Dynasts*.[31] The omission conforms to his overall portrayal of Sir John. While he dramatizes Nelson's every step—focusing on his refusal to remove his stars and orders or to cover them with a great coat, the blow of the musket ball that kills him, his stoicism at the approach of death—Hardy consigns Moore's rally of the Forty-second to a "Dumb Show" and narrates his wounding through the dialogue of two stragglers who share rumors.

Hardy distances us from Moore's heroism, reverting to dramatic treatment only to capture the firmness with which he, like Nelson, faces death. The way Hardy links, yet distinguishes, Moore and Nelson shows how any writer, whether artist or historian, inevitably manipulates emphasis to supply, in Carlyle's phrase, not the real thing but some "plausible scheme and theory" of the real thing. How history as plausible scheme and theory might translate into a narrative construct was suggested to Hardy by T. H. Green's *Introduction to Hume* (1874), in which he found the statement that historical method consists of rationalizing the "chaos of events" by tracing through it "a connected series of ruling actions plus beliefs."[32]

A connected series of ruling actions describes the plot of *The Dynasts*. Hardy, like White's modern scientist, seems to reject teleology for contiguity—but only seems. The artist's hand, guided by beliefs, is throughout fashioning order from the chaos of events. To cite an especially telling instance, when Pitt, having received news of Austerlitz, directs Wiltshire's servant to roll up the map of Europe because " 'Twill not be needed now / These ten years!" (I, vi, vi, 128), he is precisely right. Hardy, to stress this moment, sacrifices history to drama and condenses two events into one. Though his sources (Earl Stanhope, Lord Rosebery, and W. M. Sloane) agree that Pitt made his prophetic remark not on reading the courier's dispatch at Shockerwick House but on returning home to Putney, Hardy uses it as an ironic curtain speech to the minister's tour of Wiltshire's picture gallery.[33]

For Pitt, the tour appears to offer what George III, in their interview at Gloucester Lodge, had prescribed: respite from the "strains of office" (I, vi, i, 63). Wiltshire, alerted by Pitt to the beat of an approaching horse, hopes it brings "no ugly European news / To stop the honour of this visit" (I, vi, vi, 127). That ugly European news is what the horse brings we, witnesses to the allied debacle at Austerlitz, are already aware. From our perch, even more Olympian than that of the Spirits, we detect in *The Dynasts'* connected series of ruling actions a sustained dramatic irony. In Pitt's comment after lingering before Gainsborough's portrait of the actor Quin (" 'Nature, in spite of all his skill, crept in; / Horatio, Dorax, Falstaff—still 'twas Quin' " [I, vi, vi, 126]), we recognize a foreshadowing of the minister's fate. Despite art, life creeps in. Neither the elegance of Wiltshire's pictures nor Pitt's diplomatic skill shields him from Europe's ugliness.

Hardy depicts Pitt almost instinctively acknowledging that his contrivances cannot so shield him. He peruses Wiltshire's collection, one ear to the road, hearing the horse before it becomes audible to his host. In the courier's news he reads his own end. To his question "Where is Austerlitz?" he himself replies, "What corpse is curious on the longitude / And situation of his cemetery! ... I am as though I had never been!" (I, vi, vi, 127–28). Watching by Pitt's deathbed some weeks later, Bishop Tomline identifies his disease as "Austerlitz" (I, vi, viii, 134). Austerlitz is Europe's disease too. That incurious corpse is not only Pitt but the prone, suffering human form that Years reveals to his fellow Spirits, configured by the geography, and Wiltshire's map, of the Continent.

Pitt's death, and the death of old Europe, close Part First. His decline parallels—is enmeshed in—Napoleon's rise. His last words—"My country! How I leave my country!" (I, vi, viii, 136)—echo Nelson's and anticipate Moore's, but to very different effect. Nelson knows he leaves his country stronger for his sacrifice and that he will be judged accordingly. While Moore knows he leaves his country stronger, he does not know that his sacrifice will be so judged. Pitt knows neither that he leaves his country stronger nor that his countrymen will forgive his failure.

Portrayed as the victim of George III's madness, Hardy's Pitt epitomizes the irony of the world picture *The Dynasts* projects. He is welcomed by George to a spa in South Wessex where he can rest but is deprived of rest by George's unwillingness to welcome Pitt's proposed coalition with Fox. The king represents himself as a rational man, who bears "in common" with Pitt the strains of office (I, iv, i, 63), but reacts irrationally, increasing those strains on his minister, ensuring the succession of the party he loathes, and threatening to infect the country with a chaos comparable to the chaos within him: "Rather than Fox, why, give me civil war!" (I, iv, i, 67).

Yet, when confined because of illness in 1811, the king so deftly punctures Sir Henry Halford's announcement of the victory at Albuera (the bloodiest battle of the Peninsular campaign) that one of his attendants is moved to remark, " 'Twould seem / the madman were the sanest here!"

> He says I have won a battle? But I thought
> I was a poor afflicted captive here,
> In the darkness lingering out my lonely days,
> Beset with terror of these myrmidons
> That suck my blood like vampires! . . .
>
> . . . .
>
> When will the speech of the world accord with truth,
> And men's tongues roll sincerely!
>
> (II, vi, v, 305–6)

George perceives in the doctors applying their barbarous remedies myrmidons who ravage his body, as other myrmidons, red-coated and blue, ravage Spain and one another. His is a world gone awry, where kings learn wisdom only in madness, healing can kill, and reason's tongues roll insincerely.

Reason's insincerity is epitomized by Napoleon. Sprung from reason's most colossal public triumph, the French Revolution, he betrays the Revolution. As the Spirit of the Pities whispers after Napoleon has crowned himself in Milan Cathedral, "Lieutenant Bonaparte, / Would it not seemlier be to shut thy heart / To these unhealthy splendours?—helmet thee / For her thou swar'st-to first, fair Liberty?" (I, i, vi, 35). But Lieutenant Bonaparte would succeed Louis XIV as the Sun King. His entrance has, the archbishop flatters him, "streamed radiance on our ancient capital" (p. 33).

That radiance reappears when the sun breaks forth "radiantly" (I, vi, iii, 116) at Austerlitz. Though the sun of Austerlitz was to become the emblem of Napoleon's advance eastward, Hardy overlays it with an ambivalence beyond his Emperor's perception. The sun rising at Austerlitz reveals Napoleon's face as "ash-hued" (p. 117)—deathlike. As *The Dynasts* unfolds, the image of the sun accrues increasingly ironic (and malignant) implications: from the midsummer day, darkened by storm clouds, when the Grand Army crosses the Niemen into Russia; to the sun setting in the faces of the French on the eve of Borodino, then rising red the next morning; to its parody, the "lurid, malignant star" (III, i, viii, 348) of fire above the Kremlin; to the dawn of no sun, the rainy sky at Waterloo, and the sun's disappearance below the horizon as the French retreat becomes a rout; to Napoleon's image of himself as a meteor burned-out in the Wood of Bossu.

Hardy's Napoleon consorts with death like the carrier of a lethal disease. When he crosses the Niemen, he hums "Malbrough s'en va-t-en guerre," to which the Spirit Sinister adds an appropriate coda: "*Monsieur d'Malbrough est mort, / Est mort et enterré!*" (III, i, ii, 330–31). The Emperor sings "Marlbrough s'en va-t-en guerre" again as he waits to cross another river after the disaster at Leipzig. Hardy took this curious detail from Sloane, incorporating it into an elaborate symbolism introduced by Napoleon's command (the first crucial decision we see him make) to forgo England and turn east.[34] Whereupon, from his eminence above the St. Omer Road, he watches his army recede "till each column is seen but as a train of dust" (I, iii, iii, 63).

This image foreshadows the whole drama, which largely consists of armies on both sides, at Napoleon's command, receding into dust. Captain Hardy tries to comfort Nelson, as he complains of his fading sight, by attributing it to "smoke from ships upon our win'ard side, / And the dust raised by their worm-eaten hulks"—to which Nelson replies, ". . . 'tis dust of death / That darkens me" (I, v, iv, 94). Dust, both of nature and of death, pervades the air of Salamanca, where "fogs of dust" (III, i, iii, 336), kicked up by French and English cavalry, recall earlier fogs: "the kind shroud," as Napoleon styles it (I, vi, iii, 116), under which the French move to the assault at Austerlitz; the miasma into which the English fade at Walcheren. The shroud of Austerlitz ironically suggests the state of the Grand

Army entering Moscow, reduced to one-fourth its original strength; or retreating across Lithuania, "enghosted [another of Hardy's puns] by the caressing snow" (III, i, ix, 354); or at last, as Napoleon must explain to Marie Louise, "gone all to nothing" (III, i, xii, 363).

While we do not quite follow Napoleon all to nothing, Hardy asks us to fill in the blank. Fueling rumors of the impending Franco-Russian War (it is midsummer, 1811), Years prophesies to Prime Minister Spencer Perceval that "*the rawest Dynast ... Will ... Down-topple to the dust like soldier Saul*" (II, vi, vii, 320). Years's prophecy essentially closes Part Second and prepares us for Part Third, the play's denouement. We know, though Perceval cannot, that the prophecy of the pale, hollow-eyed gentleman at the Prince Regent's gala will be fulfilled. Promising the event "*before five more have joined the shotten years*" (p. 320), the Spirit seems to refer solely to Napoleon's final defeat. But the image of dust invites us to read in the Emperor's occupation gone his death as well.

Hardy's Napoleon himself senses that death is his constant companion. It is because at his back he always hears time's winged chariot that he divorces Josephine to wed Marie Louise. When he tells Josephine that the empire requires her sacrifice, she answers, "I hoped and hoped the ugly spectre [their failure to produce an heir] / Had been laid dead and still" (II, vi, ii, 260). But for Napoleon, the specter also embraces "the unexpected, lurid death of Lannes ... Tiptoed Assassination haunting round / ... the near success / Of Staps the madman" (II, vi, i, 253–54).

That the threat Napoleon perceives in these apparitions weighs on his private self is underscored by the fear he acknowledges feeling about them in a speech Hardy labels a soliloquy, and by their re-emergence in the nightmares Napoleon (like an earlier, Shakespearean tyrant) suffers—of the risen, reproachful corpses led by the Duke of Enghien; and of a mutilated Marshal Lannes rebuking him, " 'What—blood again? ... Still blood?' " (II, vii, vi, 501)—before Ligny and Quatre-Bras and midway through Waterloo. Whether Napoleon's dream of Enghien is biography or invention, Hardy found Lannes's indictment of the Emperor in Pierre Lanfrey's account of the Marshal's agonizing end at Wagram: "He burst forth into bitter complaints against the ambition and insensitivity of the reckless gambler, in whose eyes men were but so much ready corn, to be risked without scruple and lost without remorse."[35]

By incorporating into Napoleon's psychic life an encounter that even Lanfrey concedes to have been only rumored, Hardy probes his emperor's mind to a depth that the boldest of nineteenth-century historians would not have attempted. Napoleon's dreams and the ghastly figures that haunt his waking thoughts image his inmost fears. They also reveal him as more complex than the ruthless cynic Lanfrey takes him to be. Those faces and voices that rise to accuse him come from within himself. Tormented by the pain he causes, he feels helpless to stop it. "Why," he asks, confronted by his dream of Enghien, "why should this reproach be dealt me now? / Why hold me my own master, if I be / Ruled by the pitiless Planet of Destiny?" (III, vi, ii, 468). His denial of responsibility for his actions echoes his deflection of Queen Louisa's plea for the return of Magdeburg: "My star is what's to blame—not I"; and this disclaimer elicits Years's remark that Napoleon is one *"of the few in Europe who discern / The working of the Will"* (II, i, viii, 179).

The working of the Will seems a function both of the mind and of the impenetrable mystery beyond. Napoleon's soliloquy, in which he catalogues the terrors that warn him to "send down shoots to future time" (II, vi, i, 245), is addressed to the Spirit of the Pities as well as to himself. The general chorus of the Spirits, hovering by Pitt's deathbed while Years recalls having long *"communed with that intelligence,"* pronounces its efforts to understand history *"but the flower of Man's intelligence,"* which it then dismisses as *"an unreckoned incident / Of the all-urging Will"* (I, vi, viii, 137).

Showing Years in communion with a human intelligence to impart an edict of the Will is also how Hardy dramatizes Villeneuve's decision to commit suicide:

SPIRIT OF THE YEARS

I'll speak. His mood is ripe for such a parle.
(Sending a voice into VILLENEUVE's ear.)
Thou dost divine the hour!

VILLENEUVE

But those stern Nays,
That heretofore were audible to me
At each unhappy time I strove to pass?

SPIRIT OF THE YEARS

Have been annulled. The Will grants exit freely;
Yea, It says "Now." Therefore make now thy time.

(I, v, vi, 105)

Hardy meshes the voices in this dialogue as if to blend Years
into Villeneuve's consciousness. Years acknowledges that
Villeneuve's state of mind is what opens him to the Will's urging. The admiral's reminder of the "Nays," which earlier for-
bade him to act on his wish for death, metrically completes the
blank verse line begun with Years's announcement that the
hour has arrived. The reminder, though interrogatively put, is
then turned by Years into a declarative statement ("Those stern
Nays . . . Have been annulled"), in effect superimposing the Immanent Will on an individual will.

The Stuff of Myth

Hardy's treatment of Villeneuve's suicide displays a rhetorical
subtlety with which he is seldom credited. It dramatizes the
tension, pervading *The Dynasts*, between history as a series of
discrete events, experienced or observed, and the unfathomable
mystery that frustrates the search for meaning in their substance and sequence: the tension between the apparently arbitrary course of history and the artist's efforts to impose order
on it. If all incidents are, as the Spirits in chorus assert,
"unreckoned," and if "the flower of man's intelligence" is
among those unreckoned incidents, life is absurd and artists,
who claim to find order, significance in life, are deluded—
which is how the Spirit Ironic sees it. At Talavera, he shrugs
off the spontaneous friendship between French and English soldiers, who drink and clasp hands across the Alberche Brook, as
"Life's queer mechanics chancing to work out in this gro-
tesque shape just now" (II, iv, vi, 245). In Hardy's version, however, the engagement at Talavera appears not grotesque but

symmetrical. The main battle commences at dawn on July 28, 1809, and ends at dusk. The troops united by sustaining water, *"sealing their sameness as earth's sojourners"* (p. 245) under the glare of noon, are united again by consuming fire, *"opposed, opposers, in a common plight"* (p. 246), in the conflagration which sweeps the field during the night. The balanced construct Hardy makes of the Battle of Talavera illustrates how he transcends his sources for symbolic purposes. Napier, who fought on the Peninsula, but whose English bias is softened by respect for the skill and courage of French soldiery, barely touches on the brief truce at Alberche Brook. Louis Adolphe Thiers, whose loyalty to the Revolution and Napoleon (until his Russian misadventure) persuades him to equate England with villainy, ignores it. Having no impact on the outcome of the battle, it has, for neither historian, any importance.

But for Hardy, the incident at Alberche Brook exemplifies the commonality of all men (the kind of irony to which the Spirit of the Pities, rather than the Spirit Ironic, responds) and becomes a vehicle to image the nearly apocalyptic scope the Napoleonic Wars attained. Hands clasped across the stream yield to the slaughter of English ranks by French cavalry, *"Who ring around these luckless Islanders, / And sweep them down like reeds by the river-brink / In scouring floods; till scarce a man remains"* (p. 246). The water the French and English have shared metamorphoses, in the simile of men swept down like reeds by scouring floods, into an image of massacre: rhetorically, water and fire are made to conjoin. Their conjunction recurs at Salamanca, where the Spirit of the Pities follows Wellington's columns, *"Still headily [pursuing] their way, / Though flood and foe confront them, and / The skies fling flame"* (III, i, ii, 334); and in the Russian wasteland, where Kutuzov and his staff find French bodies *"cindered by the fire in front, / While at their back the frost has caked them hard"* (III, i, xi, 359).

Hardy may have taken a hint for this conjunction from John Holland Rose, who concludes his account of the remnants of the Grand Army in flight across the Berezina by attributing to the men trapped between the Russians and the burning bridge a "despair or a loathing of life [which] drove many to end their miseries in the river or in the flames."[36] Whatever Hardy's debt to his historian mentors, the yoking of fire and water imposes a mythic dimension on events. It suggests that history and myth are composed of the same stuff.

Tolkien also believes that history and myth are composed of the same stuff. He perceives in history the promise of a Christian "eucatastrophe" (a somewhat odd term he coined to define the sudden, joyous reversal that ends most fairy tales).[37] And though real catastrophe (in Tolkien's equally odd term "dyscatastrophe," sorrow and failure) seems the more likely promise of history, as Hardy presents it through the agnostic frame of *The Dynasts*, his use of fire and water in their apocalyptic acceptations has Christian roots. "I . . . baptize you with water unto repentance," John the Baptist warns his congregants in the wilderness of Judea, "but he that cometh after me . . . shall baptize you with the Holy Ghost and with fire: . . . he will thoroughly purge his floor, and gather his wheat into the garner; but he will burn up the chaff with unquenchable fire" (Matthew 3:11–12). *The Dynasts* dramatizes a conflict that appears to all but the English (at least until Napoleon's debacle in Russia, and then once more after his escape from Elba) an unquenchable fire burning both wheat and chaff. In the French and Prussian armies locked in combat at Quatre-Bras, the Spirit of the Pities sees "an unnatural Monster, loosely jointed, / With an Apocalyptic Being's shape"—and Years agrees: "Thou dost indeed. / It is the Monster Devastation" (III, vi, vi, 474).[38] Hardy's Europe is like the Europe Blake, mythicizing a prior phase of the same history, envisions as forests of eternal death, where kings devour and are devoured, consume and are consumed.

Kings devouring and devoured, their subjects consumed and consuming, epitomize the action of *The Dynasts*. Alexander offers up his sister, Ann; Francis sacrifices his daughter, Maria Louisa. Napoleon sacrifices Josephine. He essentially consumes Villeneuve, as George III consumes Pitt. Finally, Napoleon consumes himself. Hardy dramatizes this struggle of his kings to devour their rivals—in the process of which they devour loyal subjects and even their own kin—by pairing their victims. Josephine and Maria Louisa are counterparts. Villeneuve finds his counterpart in Nelson. Napoleon finds his initially in Pitt, eventually in Wellington.

When Francis insists to Metternich that Maria Louisa be allowed herself to decide whether to accept Napoleon's proposal of marriage, the chancellor comments, "So much for form's sake! . . . What she must do she will; naught else at all" (II, vi, iii, 269). He is right. Raised in the dynastic ethos, Maria Louisa replies, "My wish is what my duty bids me wish" (II, v, iv, 272). But duty in Maria Louisa's sense, which demands she wed a man she has been taught to hate, cannot reconcile the bourgeois Josephine to a divorce from a man she still loves. She believes in the sanctity of marriage and appeals to Napoleon "by [their] old, old love, / By [her] devotion" (II, v, ii, 261)—feelings irrelevant to the game of dynasties. When Maria Louisa adduces their opposites ("But, Chancellor, think what things I have said of him!") as obstacles to wedding Napoleon, Metternich dismisses them as "words" (II, v, iii, 271).

Hardy stresses this reduction of things to words through the parallel syntax of the dialogue. Maria Louisa's allusion to the "things" she has said of Napoleon provokes Metternich's succinct question, "Things?" which is echoed by her elaboration, "Horrible things," prompting his deflating conclusion, "Words." The denial of reality to things and the implication that dynasts can have things be as they would have them be epitomize Bonaparte. He challenges English naval power despite Villeneuve's warnings; explains to Josephine that divorce need not mean a total parting; calls on heaven, with Moscow burning around him, to "curse the author of this war" (III, i, viii, 351); persuades himself at Leipzig, to the discomfiture of his marshals, that he can dupe the allies by deploying his outnumbered infantry in two lines instead of three, and at Waterloo that he can snatch victory from defeat by telling his exhausted troops that the fresh columns entering the field are Grouchy's rather than Blücher's.

By such deceptions, he destroys Villeneuve's fleet, the Grand Army, himself. Suborning the troops sent to intercept him at La Mure, on his return from Elba, Napoleon is assured by a grenadier, "You are the Angel of the Lord to us" (III, v, iv, 437). His image compellingly suggests Blake's Edward III, who spreads war's "desolating wing" (1: 2, 48; E., p. 416) over France. And in *The Dynasts* the image recurs. About to leave the Duke of Richmond's ball for the field at Quatre-Bras, the doomed Brunswick is "stirred by inner words [actually Years's forebodings], / As 'twere my father's angel calling me" (III, vi, ii, 459).

Hardy, who surely never read *King Edward the Third*, drew this incident from Henry Houssaye's otherwise unimaginative *1815: Waterloo*, which records "Brunswick, through a sort of presentiment, [feeling] the shudder of death."[39] Sir Thomas Picton also feels the shudder of death. Shadowed, like Brunswick, by an apparition of Years's "tipstaff," he *"cooly conned and drily spoke to it"* (III, vi, iii, 463).

Before Waterloo Picton boasts (a remark he is said actually to have made), "When you shall hear of *my* death, mark my words, / You'll hear of a bloody day!" (III, vii, iv, 493). His imperturbability contrasts with the melancholy of Brunswick, who (Hamlet-like) wears black in memory of his father, dead nine years, and of whom Wellington observes, "He is of those brave men who danger see, / And seeing front it,—not of those, less brave / But counted more, who face it sightlessly" (III, vi, ii, 460).

Wellington's appreciation of Brunswick is one of the virtues Hardy dramatizes to contrast the Duke with, and assert his superiority to, Napoleon in the conflict of mighty opposites (suggestive of Scott's in his *Life*) that occupies the last act of *The Dynasts*. For Napoleon also has a kind of Brunswick in Villeneuve, who, with his rival, Nelson, manifests a crucial tension between the reflective man and the man of action. When Napoleon is angered by Villeneuve's failure to appear in the Channel, his Minister of Marine Decrès explains that the admiral's "drawback [is] that he sees too far" (I, iii, i, 58). Napoleon, absorbed in his own ego and cavalier about the details of naval warfare, suffers no such drawback: "My brain has only one wish—to succeed" (p. 59). He confesses himself akin to Decrès' image of Nelson: "A headstrong blindness to contingencies / Carries [him] on, and serves him well / In some nice issues clearer sight would mar" (p. 59). In delineating this contrast between Villeneuve, who thinks too much, and Nelson, who single-mindedly plunges ahead, Decrès exaggerates, but only somewhat. The attitude that he disparages as peculiarly British (although it is partly shared by Napoleon)—"fatuous faith in one's own star . . . Smugly disguised as . . . trust in

God" (p. 58)—emerges at Trafalgar in Nelson's pronouncement, "We must henceforth / Trust to the Great Disposer of
events, / And justice of our cause!" (I, v, ii, 85). Despite warning
his second-in-command Collingwood that he expects "tough
work against the French fleet" (I, ii, i, 39), the prospect of
"black outcomes" (p. 40), which haunt Villeneuve and which
he broaches in his letter to Decrès, never troubles Nelson.

That the rival admirals should be perceived as bipolar is im-
plied by Hardy's transition from the meeting of Nelson and
Collingwood (I, ii, i) to the arrival, while Villeneuve is penning
his misgivings to Decrès, of General Lauriston with Napoleon's orders to his fleet (I, ii, ii). The sunshine "of those dear
Naples and Palermo days" (I, ii, i, 40), the memory of which
Nelson cherishes, is eclipsed by the black outcomes Villeneuve
predicts for France's navy. Nelson's "ill-timed confessions" (p.
40) about his personal life, for which he apologizes to Collingwood, give way to the even more ill-timed confession Villeneuve sends Decrès.

This Shakespearean dovetailing of scenes enables Hardy to
highlight the psychological, and partly irrational, forces in history, and hints at an ironic dimension of Napoleon's fall. As-
sailed by the same doubts that tormented Villeneuve, he must
face in Wellington a Nelson with none of Nelson's reason for
self-recrimination and none of his Villeneuve-like longing for
death. Waterloo thus balances Trafalgar. The drama turns full
circle. The ending of *The Dynasts* projects a Europe restored to
an old order that the Revolution had seemingly shattered beyond the power of kings' horses and men to put together again.
But structurally Part Third breaks the six-act pattern of Parts
First and Second. Waterloo stands alone as a seventh act. Hardy
has adapted a device from another great, unstageable drama,
Shelley's *Prometheus Unbound*, the fourth act of which forms
an apocalyptic coda to the action proper.

While no one would term Waterloo a coda to the action of
The Dynasts, it comes as close to being an apocalyptic event as
Hardy's agnostic and open-ended view of history would allow
him to contemplate, as its departure from the symmetry of the
whole is meant to stress. Jean R. Brooks has summarized the
symmetrical features of *The Dynasts*, among them the parallels between Napoleon and Wellington: both forty-six when

Waterloo was fought, each riding a horse with a name that suggested epic grandeur, neither able at the crisis to reinforce his troops.[40] She argues (I think oversimplifying) that Part Third entwines Napoleon's fall and Wellington's rise—as Part First entwines Pitt's fall and Napoleon's rise. Though the Emperor's increasing corpulence and somnolence, his decreasing hold on reality, measure his deterioration, its germ is latent in his conduct from the outset. As his ill-treatment of Villeneuve has driven the admiral recklessly to challenge Nelson's fleet at Trafalgar, his ill-treatment of Marshal Ney drives him recklessly to charge Wellington's squares on Mont Saint-Jean. And watching Ney all but pierce the English line, yet lacking the reserves to support him, Napoleon recognizes what Villeneuve had recognized when he had bent to his Emperor's commands: "Life's curse begins, I see, / With helplessness!" (III, vii, vii, 504).

Brooks would have us conclude that Wellington escapes life's curse, that Part Third of *The Dynasts* depicts him gaining control as Napoleon loses it. But the Duke has no counsel to offer his men except to "fall upon the spot we occupy, / Our wounds in front" (III, vii, vii, 507). Waterloo is, for Hardy, a soldier's battle. The English win because, in Wellington's words, they "pound the longest" (VII, vii, iv, 497). Pitt's tribute to his countrymen after Trafalgar might have been uttered just as appropriately after Waterloo. As Napoleon concedes, every European dynast has bowed to him except "always England's . . . / Whose tough, enisled, self-centered, kindless craft / Has tracked me, springed me, thumbed me by the throat, / And made herself the means of mangling me!" (III, vii, iv, 520).

Despite its insistence on the ultimate mystery of things, *The Dynasts* is a drama of England saving Europe by her example. Hardy imposes literary form on events for which no sufficient intrinsic rationale can be discerned. He closes up time and traverses space, turning human struggle perhaps not into history, as historians define it, but into a kind of mytho-history. Which is not to dismiss *The Dynasts* as history. When he recounts the collapse of the Old Guard in the gathering dusk of June 18, 1815, John Keegan builds this last stage in the Battle of Waterloo—as Hardy might well have—into a symbol. More than an end to Napoleon's career, it is, for Keegan, the real end of the French Revolution:

the reduction of the Guard to a fugitive crowd was also the reversal of the most powerful current in recent European history. The Revolution had made itself manifest by the Parisian crowd's defeat or subversion of the royal army in July, 1789; the metamorphosis of the Guard into a crowd, its only motive self-preservation, its only purpose flight, marked, as effectively as anything else we can point to, the restitution of power to its former owners.[41]

The fabric into which Hardy weaves the events of the Napoleonic Wars is the same kind of fabric into which writers of narrative—historians as well as novelists and poets—recurrently weave experience. Without such transactions, between the writer and his subject, between the reader and that subject crafted by the writer, neither fiction nor history would make sense.

Conclusion
(Mainly about
Conclusions)

When John Keegan asserts that the rout of the Old Guard at Waterloo marks the real end of the French Revolution, he treats the scene as a novelist might: to epitomize the conflict by which the Revolution, having displaced the ancien regime, is ultimately subverted and the ancien regime itself restored—the process by which, as the Hardy of *The Dynasts* would have insisted, history comes full circle. Historians of the Revolution would doubtless argue that, however dramatic, Keegan's image of the Guard as revolutionary France broken at last veneers a tangle of forces to be unraveled and reordered only by systematic inquiry. And a myriad of books exist that attest to this view. But even practitioners of systematic inquiry who differ as much from Keegan, and from each other, as Lord Acton and Georges Lefebvre are drawn to certain events of the Revolution (as Keegan is to the collapse of French arms on Mont Saint-Jean), because they find them rich with symbolic import. Acton describes the rebuff by Mirabeau and Bailly, in June 1789, of Louis XVI's messenger, de Brézé, who carried the royal command that the Third Estate disperse (a scene that makes a nice counterpoint to the scene unfolding a quarter century later on Mont Saint-Jean), and concludes, "It was at that moment that the old order changed and made place for new." Lefebvre describes the army's overthrow of republican institutions on eighteenth Brumaire (another scene that might be considered a counterpoint to the scene on Mont Saint-Jean) and concludes,

"Bonaparte [the military dictator] put Sieyès [the constitutionalist] into eclipse."[1]

Behind these assertions, presented simply as fact, lies a muted allegory, in which parts become microcosms of the whole and personalities bespeak parties, passions, and ideologies at war. The histories through which they parade are, consciously or unconsciously, shaped to reflect the convictions of their historians. Keegan, a lecturer at Sandhurst, dates his *finis* to the Revolution 1815 when, as he puts it, power in France (essentially the army) reverted to its former owners. Lefebvre, a Marxist, dates his *finis* to the Revolution 1795 when, the Sanscullottes having been starved into submission, the bourgeoisie solidified their claim to ownership: "This [Prairial of the Year III or May 1795] is the date which should be taken as the end of the Revolution. Its mainspring was now broken."[2]

Endings, then, function as symbolic statements, devices that project the constructions historians place on history. Barbara Herrnstein Smith observes that effective closure in a poem (or in this case a play) renders whatever might follow irrelevant (Hamlet is dead), or predictable (Fortinbras restores order, Horatio recounts Hamlet's ordeal), or "another story" (Horatio has further adventures).[3] Much the same observations might be made of historical narratives, whether purportedly factual, like the biographies of Napoleon by Scott and Hazlitt, or transparently fictional, like Dickens' *Tale of Two Cities*. What happens after the last page is irrelevant (Napoleon and Carton are dead), or predictable (the Bourbons resume the crown, the Darnays live happily ever after), or another story (the revolutions of 1830 and 1848 are yet to come).

Smith also explores the consequences for poems (again in observations that pertain to historical narratives) when the poet either fails to achieve, or deliberately avoids, effective closure: thus we are left aware of time ongoing and of characters unfulfilled, their situations unresolved, because the themes they act out remain ambiguous, the narrative structures their actions comprise remain open.[4] Do we really need, as Scott and Hazlitt suppose, to follow Napoleon to St. Helena, there to witness the sordid details of his dying? Is not the Emperor dead to history, as Hardy in *The Dynasts* and Lefebvre in his two-volume study *Napoleon, from 18 Brumaire to Tilsit* and *Napoleon, from Til-*

sit to Waterloo imply, the moment his lines break on Mont Saint-Jean? Aesthetically, symbolically, Hardy and Lefebvre are surely right. The long accounts of Napoleon's exile with which Scott and Hazlitt close their biographies seem no more than extended anticlimaxes. Both Scott and Hazlitt have, however, ideological aims sufficiently important to them to override the demands of artistic structure: Scott to vindicate, Hazlitt to calumniate the conduct (proper or perfidious?) of Albion toward the fallen Emperor.

This debate concerning Napoleon's demise reflects a notion first proposed by Robert M. Adams and repeated by J. Hillis Miller: that in stories, death eludes finality. Miller calls death the most open-ended of endings.[5] The blurring of finality, even in death, has to do with the special vantage the writer enjoys, and shares with his reader, particularly when the subject is history. W. Wolfgang Holdheim ascribes what he perceives as the teleological character of Aristotelian plots to the likelihood that their authors began with endings already in mind.[6] Though the work habits of Dickens and Tolkien, for two, suggest that Holdheim may overgeneralize about how authors begin, the position authors occupy, outside the actions they narrate, grants them the possibility of omniscience, enables them to see things whole. Indeed, if the actions they narrate are fact rather than fiction, they can see beyond the whole they construct to the untold tales that happened before and will happen after.

While history provides prefabricated materials for the manufacture of plots, it also poses problems of control in their design that fiction may not. Carlyle, as John D. Rosenberg points out, wrestles with this dilemma both in "On History" and in *The French Revolution*. Underlying the gap he concedes in "On History," between the linearity of narrative and the solidity of action is his awareness that beginnings and endings are, in Rosenberg's phrase, "chimeras of the mind."[7] He thus ends his narrative of the French Revolution with the emergence of Napoleon as its champion in 1795, which to Rosenberg seems so arbitrary a choice of ending as to constitute no ending at all. He interprets it as Carlyle losing interest as Napoleon gains influence.[8]

But to read the ending of Carlyle's *French Revolution* as an expression of his loss of interest is to trivialize a strategy of closure that nicely completes the narrative's symbolic scheme. With the emergence of Napoleon, the Revolution has come full circle: from the breakdown of order attendant on the collapse of the monarchy, to the chaos of the Terror, then to a resumption of order (however temporary) embodied in the Emperor-to-be. Carlyle views the Revolution, and implicitly the Napoleonic era, from a temporal perspective similar to Tocqueville's and reaches a similar conclusion. For Tocqueville, the significance of Napoleon's rise to power is at least in part that it restores the French monarchy.[9] For Carlyle, Napoleon's rise is akin to Cromwell's. Characterizing each in *On Heroes and Hero-Worship* as in his own way "the necessary finish of a Sansculottism," he declares that "the manner in which Kings were made, and Kingship itself took rise, is again exhibited in the history of these Two" (5: 204).

Tocqueville openly acknowledges that he structured his story of the Revolution to suit his thesis that the new regime was the old in altered dress: "I shall stop at the point at which the Revolution appears to me to have, to all intents and purposes, achieved its aim and given birth to the new social order, and then proceed to examine the nature of this social order in some detail."[10] His examination of the new social order—imposed, he argues, by a government more centralized than any since that of ancient Rome—yields a portrait of France essentially (if not evaluatively) like the portrait Marx draws, in *The Eighteenth Brumaire of Louis Bonaparte*, of France enmeshed in a bureaucracy imaged as an "appalling parasitic body," a Blakean Polypus, "[choking] all its pores" (p. 121).

Marx perceives the Revolution ending, however, not with Napoleon's assumption of power in 1800 but with his abdication in 1814. For Marx, events proceed in dialectical spasms. They are triggered first by such men as Camille Desmoulins, Danton, Robespierre, and Saint-Just, who, having "knocked the feudal basis to pieces and mowed off the feudal heads," struggled toward republican (Marx's word is "bourgeois") institutions; and they are then redirected by Napoleon. Though he dismantled republican institutions, he "created inside France the conditions under which alone free competition could be developed . . . and everywhere swept the feudal institutions away, so far as was necessary to furnish bourgeois society in France with a suitable up-to-date environment on the European continent" (p. 16).

Napoleon is (for Lefebvre as well as for Marx) an instrument of bourgeois aspirations, and thus not a contradiction to the Revolution but its completion—at least, Lefebvre specifies, "in terms of its professed aims of 1789."[11] Lefebvre's phrase "professed aims" reflects a predilection he shares with Marx for the kind of metaphor Collingwood and, to different purpose, Darwin would reject because it attributes personality, consciousness, intention to impersonal agents. Citing the civil wars that shook Britain in the seventeenth century ("the first two modern revolutions") as precursors to 1789, Lefebvre discerns in them the English bourgeoisie's campaign, "in the guise of" religious conflict, to seize economic power.[12] He is echoing Marx, who discovers in Cromwell and his followers a movement that "borrowed speech, passions and illusions from the Old Testament" but was ultimately to discard them for the speech, passions, and illusions of empiricism: "When the real aim had been achieved, when the bourgeois transformation of English society had been accomplished, Locke supplanted Habakkuk" (p. 17).

Marx represents the French Revolution in a comparable way. In effecting their own transformation of a feudal into a bourgeois society, the heroes of the Revolution and their followers "performed," as Marx puts it, "the task of their time" (p. 16). His phrases "real aim" and "the task of their time" suggest, if only metaphorically, a Hardy-like Will that deflects people's endeavors toward aims beyond those they profess and determines the course of events. He places his faith in revolution, its promise of a utopian future, as Pascal placed his faith in revelation.

This analogy is proposed by Martin Jay, who borrowed it from Lucien Goldman, and uses it to characterize Marxism as "arguably" a religion, though one drained of transcendental values.[13] Jay's characterization of Marxism as a religion (with God left out) is hardly unique. M. H. Abrams finds in Marx's idea of history the same dialectically driven spiral posited by, say, Blake or Coleridge—and an insistence even stronger than theirs on the Biblical vision of an end punctuated by apocalyptic violence.[14] What Marx labels the first French Revolution "moves along an ascending line": the absolute monarchy overthrown by the Constitutionalists overthrown by the Girondins overthrown by the Jacobins. The Revolution of 1848 "moves on a descending line": the proletariat betrayed by the petty bourgeois democrats betrayed by the moneyed bourgeois republicans betrayed by the party of order betrayed by the army and Louis Bonaparte.

1848 thus completes the circle adumbrated by the famous opening of *The Eighteenth Brumaire*, in which Marx quotes Hegel's remark "that all facts and personages of great importance in world history occur . . . twice" (p. 15). Or as Carlyle analogously accounts for such recurrences in "Jesuitism," where he depicts God as the weaver of history working his shuttle, "the loud-roaring Loom of Time, with all its French revolutions, Jewish revelations, 'weaves the vesture thou seest Him by' " (20: 325–26).

History as a dialectical process, or as the snake with its tail in its mouth, is a concept of Christian no less than Marxist historians. Its movement of turn and return toward a utopian or apocalyptic end could scarcely find a more elaborate formulation than that of the assertively Christian Arnold Toynbee, who traces the dynamic behind the growth of civilizations to "an *élan* which carries them from challenge through response to further challenge," and images history as a wheel turning back to move forward.[15] This apparent coincidence between Marx's configuration of history and Christian configurations troubles latter-day Marxists, as well as the postmodernist thinkers E. D. Hirsch would presumably classify as "atheists," cognitive or otherwise. Although he adopts Marx's view that Napoleon consolidated the bourgeois revolution, Lefebvre abandons Marx's faith that history progresses by dialectical turns toward proletarian triumph, observing that the French Revolution ended with proletarian defeat. Walter Benjamin seeks to refurbish dialectics as an instrument for charting the course toward proletarian triumph—to "wrest Christian history from Christian historians"—by insisting that the pattern woven on the loom of time is both cyclical and open.[16]

Benjamin recasts in his own ideological mold the Romantic, and essentially biblical, model of history as an ascending spiral.

That even this modification is unacceptable to the newest left is affirmed by Michel Foucault, who diminishes the nineteenth-century's claim to having discovered the order of history by arguing that "it did so only on the basis of the circle," and dismisses universal history (in his phrase "total history") for "general history."[17] Foucault might as well have been directing his argument against Benjamin, who rejects Ranke precisely because he wrote a kind of "general history"—"Its method is additive"—and stresses that to write valid (for him, "materialistic") history means to construct rather than to accumulate: "Thinking involves not only the flow of thoughts, but their arrest," their crystallization "into a monad."[18] Whereas Foucault indicts cyclical form for negating time, Benjamin makes negating time the historian's goal; whereas Foucault advocates that the historian "deploy the space of a dispersion," Benjamin demands that he grasp eternity in an hour: "In this structure he recognizes the sign of a Messianic cessation of happening, or, put differently, a revolutionary chance to fight for the oppressed past."[19]

To fight for the oppressed past entails, in Benjamin's deliberately violent metaphor, "blasting" events from time's continuum and shaping them into the permanence of discrete artifacts. To deploy the space of a dispersion entails adding one event to another, in a chronicle of their apparently endless regress along time's continuum. In this sense Foucault's proposal for reconfiguring the narrative of history raises what seems to me a crucial question: if no text can record more than sequences of displacements or, as Derrida perceives them, mirrored images that reflect mirrored images, can a text really end, or begin, in any way except arbitrarily?[20] To ask that question is, in effect, to ask whether any text can claim a genuine conceptual basis, a rationalized structure; which is finally to ask whether meaningfully ordered discourse, much less historical narrative, is possible.

Every text obviously begins and ends somewhere—even if only, like *Finnegans Wake* (a great historical fiction of a sort), by re-beginning in its end. Writers fashion beginnings and endings as devices, guideposts, to alert readers to the way they would have their works read. Though practitioners of Derridean deconstruction may characterize history as "undecidable," the historian inevitably decides: not just where to start and stop but what to include and what to exclude, what to accentuate and what to subordinate. However vigorously one may dispute a historian's interpretation of events, his treatment of them gives significant form, meaning, to his narrative.

Gerald Graff argues that to claim otherwise (as, presumably, deconstructionists would) is to be caught in a paradox: that to read texts as if they existed in a philosophical void, moored to no preconceptions about reality, itself implies preconceptions about reality.[21] But the paradox appears to me even more fundamental than Graff suggests. If linguistic constructs can refer to nothing outside themselves—if the nature of language precludes its being about anything—how should we regard texts in which such generalizations, purportedly applicable to all texts, are made? For the most appropriate mode of expressing this critical position can only be silence.

To Fredric Jameson, Marxist historiography offers a way out of this dilemma because it reconciles the myth of an idealized plenitude or moment of completion (akin to prelapsarian Eden, or perhaps Derrida's "arche-trace") with the fact of a dialectical rhythm, through which history declines from plenitude, then turns back toward it.[22] Jameson's insistence that all theories of history, secular as well as sacred, require the prospect of some such plenitude, at least as a hypothetical possibility, is confirmed by Derrida, who presupposes for history a point of origin which, though it never existed, must be posited as if it had. Every theory of history, that is, needs its version of the Fall. As Derrida observes about the theory of the history of language, toward which he is building, "It must restore its absolute youth, and the purity of its origin, short of a history and a fall which would have perverted the relationship between outside [writing] and inside [speech]."[23]

To predicate a Fall is to promise redemption: to manifest the dialectic of history, its ultimately ascending spiral, at work. History has value, Kenneth Burke declares, only "as a documented way of talking about the future."[24] He might have been speaking for Marx, who writes his history of Louis Bonaparte's coup not to trace the path the Revolution has gone (the dead must be left behind to bury the dead) but to adumbrate the path it has yet to go:

It is still journeying through purgatory. It does its work methodically. By December 2, 1851, it had completed one-half of its preparatory work, it is now completing the other half. First it perfected parliamentary power, in order to be able to overthrow it. Now that it has attained this, it perfects the *executive power*, reduces it to its purest expression, isolates it, sets it up against itself as the sole target, in order to concentrate all its forces of destruction against it. And when it has done this second half of its preliminary work, Europe will leap from its seat and exultantly exclaim: Well grubbed, old mole. (p. 121)

Sartre too focuses on the path the Revolution has yet to go. Looking, and inviting us to look, through the mind's eye of Robespierre, he sees the Revolution as "a reality in process of totalization": false if it stops short of its predestined goal, true only "when it has attained its full development."[25] He shares Marx's sense that he is in the midst of the phenomenon he describes, trying to grasp it whole. As Dominick La Capra points out, the struggle the proletariat has lost in the streets, Marx continues in his book.[26] The rationale behind the Terror, which Robespierre never succeeded in articulating to the world, Sartre articulates on his behalf. He concludes what he appears to represent as his own statement of the dynamic underlying the Revolution with the assertion, "This is Robespierre *thinking*."[27]

Sartre acknowledges the need, which Owen Barfield acknowledges, as part of his own prescription for overcoming the consequences of the Fall, to submerge the perceiving self in the life being perceived.[28] Barfield is exploring the impact of Romanticism; and Sartre's approach to historical narrative, like Marx's, seems imbued with Romantic ideas. He establishes a dual perspective on history comparable, say, to the dual perspective established by Carlyle in *The French Revolution*. Distanced from events by time, both pry crucial episodes loose from its flow to make chapters in an ongoing story. Both manage also (Sartre mainly in theory, Carlyle magnificently in practice) to project themselves back into the thoughts and feelings of men embroiled in the conflict: to stand at once outside and inside their struggle.

For the struggle—revolutionary or evolutionary, toward paradise in this world or in the next—continues. Sartre and Carlyle recognize themselves to be, like all human beings who write or are written about, locked in what Frank Kermode calls the "middest." To be locked in the middest poses the writer who would comprehend history with an awesome task. If history—as not only Sartre and Carlyle but Blake, Scott, Dickens, and perhaps even Hazlitt and Hardy believe—incorporates the potential to discern where we are going in where we have been, the historian must find within himself the power of vision. He must become a prophet: Ezekiel among the captives by the River Chebar, or Blake's Bard, "Who Present, Past, & Future sees."

Notes

Introduction

1. Northrop Frye, *Fables of Identity* (New York: Harcourt, Brace and World, 1963) 55.
2. J. H. Hexter, *Doing History* (Bloomington: Indiana UP, 1971) 4, 18.
3. Northrop Frye, *Anatomy of Criticism* (Princeton: Princeton UP, 1957) 7; Frye, *Fables of Identity* 55.
4. David Hackett Fischer, *Historians' Fallacies: Toward a Logic of Historical Thought* (New York: Harper and Row, 1970) 14, 38.
5. Peter Loewenberg, *Decoding the Past* (Berkeley: U of California P, 1985) 12, 15.
6. Henry Adams, *The Education of Henry Adams*, ed. Ernest Samuels (1906; Boston: Houghton Mifflin, 1973) 92.
7. W. B. Gallie, *Philosophy and the Historical Understanding* (New York: Schocken Books, 1964) 16; Maurice Mandelbaum, *The Anatomy of Historical Knowledge* (Baltimore: Johns Hopkins UP, 1977) 26.
8. Hexter, *Doing History* 42–43.
9. Kenneth Burke, *Attitudes toward History* (1937; Berkeley: U of California P, 1984), introduction not paginated.
10. Ibid., 308; italics Burke's.
11. E. D. Hirsch, Jr., *Validity in Interpretation* (New Haven: Yale UP, 1967) 27.
12. Jay Cantor, *The Space Between: Literature and Politics* (Baltimore: Johns Hopkins UP, 1981) 51.
13. Barbara Herrnstein Smith, *On the Margins of Discourse* (Chicago: U of Chicago P, 1978) 32.
14. Hans Robert Jauss, "Literary History as a Challenge to Lit-

erary Theory," in Ralph Cohen, ed., *New Directions in Literary History* (Baltimore: Johns Hopkins UP, 1974) 14–15.

15. Fredric Jameson, *The Political Unconscious: Narrative as a Socially Symbolic Act* (Ithaca, N.Y.: Cornell UP, 1981) 35.

16. Jonathan Culler, *Structuralist Poetics* (Ithaca, N.Y.: Cornell UP, 1975) 129.

17. Hexter, *Doing History* 30.

18. John Keegan, *The Face of Battle* (Harmondsworth, Middlesex: Penguin, 1978) 45.

19. Leo Gershoy, "Some Problems of a Working Historian," in Sidney Hook, ed., *Philosophy and History: A Symposium* (New York: New York UP, 1963) 62–63.

20. Lee Benson, "On the Logic of Historical Narration," *Philosophy and History* 34.

21. Arnold makes these assertions in "The Function of Criticism at the Present Time," *The Complete Prose Works of Matthew Arnold*," ed. R. H. Super, vol. 4, *Lectures and Essays in Criticism* (Ann Arbor: U of Michigan P, 1962) 264, 265.

22. In a letter to Dorothy Wellesley dated 22 June 1938. See *The Letters of W. B. Yeats*, ed. Allan Wade (New York: Macmillan, 1965) 911.

Chapter One

1. M. H. Abrams, *Natural Supernaturalism: Tradition and Revolution in Romantic Literature* (New York: Norton, 1971) 147.

2. Ibid., 271. Abrams quotes Coleridge's letter to Cottle.

3. Louis O. Mink, "History and Fiction as Modes of Comprehension," *New Directions in Literary History* 117.

4. Frank Kermode, *The Sense of an Ending: Studies in the Theory of Fiction* (New York: Oxford UP, 1967) 7.

5. Michael Ryan, *Marxism and Deconstruction: A Critical Articulation* (Baltimore: Johns Hopkins UP, 1982) 24, 62.

6. Ibid., 24.

7. Cited by David Couzens Hoy in *The Critical Circle* (Berkeley: U of California P, 1978) 131. Heidegger's essay, entitled "The Age of the World View," appeared in *Boundary 2*, vol. 4 (1967), with translation by Marjorie Grene.

8. William Godwin, *Enquiry Concerning Political Justice and Its Influence on Morals and Happiness*, ed. F.E.L. Priestley, 3 vols. (Toronto: U of Toronto P, 1946) 1: 32, 384.

9. Ryan, *Marxism and Deconstruction* 20; Michel Foucault,

The Archaeology of Knowledge, trans. A. M. Sheridan (New York: Pantheon Books, 1972) 9.

10. In "Freud and the Scene of Writing." Gregory L. Ulmer provides an enlightening discussion of this analogy. See his *Applied Grammatology: Post(e)-Pedagogy from Jacques Derrida to Joseph Beuys* (Baltimore: Johns Hopkins UP, 1985) 76–77.

11. Jacques Derrida, *Of Grammatology,* trans. Gayatri Chakravorty Spivak (Baltimore: Johns Hopkins UP, 1974), especially pp. 35–36.

12. Hayden White, *Metahistory* (Baltimore: Johns Hopkins UP, 1973) 147–48.

13. This is one of the analogies Abrams explores in *Natural Supernaturalism* 36.

14. Kermode, *The Sense of an Ending* 8.

15. In Fritz Stern, ed., *The Varieties of History* (New York: Vintage Books, 1973) 57; translation by the editor.

16. William Gordon, D.D., *The History of the Rise, Progress and Establishment of the Independence of the United States of America* (London, 1788), preface (unpaginated).

17. The phrase "historical Drama" is Hardy's. See Florence Emily Hardy, *The Life of Thomas Hardy* (London: Macmillan, 1962) 148.

18. Frances A. Yates, *Theatre of the World* (Chicago: U of Chicago P, 1969) 165, 167.

19. Paul Fussell, *The Great War and Modern Memory* (New York: Oxford UP, 1975) 192.

20. R. D. Laing, *The Divided Self* (New York: Random House, 1960) 117.

21. Masao Miyoshi, *The Divided Self: A Perspective on the Literature of the Victorians* (New York: New York UP, 1969) 5.

22. Edmund Burke, *A Philosophical Enquiry into the Origin of our Ideas of the Sublime and Beautiful,* ed. J. T. Boulton (1756; London: Routledge and Kegan Paul, 1958) 47.

23. As Ronald Paulson and Marilyn Butler, from their divergent perspectives, emphasize. See Paulson, *Representations of Revolution (1789–1820)* (New Haven: Yale UP, 1983) 47; and the "introductory essay" to Marilyn Butler, ed. *Burke, Paine, Godwin, and the Revolution Controversy* (Cambridge, England: Cambridge UP, 1984) 1–2.

24. Herbert Lindenberger, *Historical Drama: The Relation of Literature and Reality* (Chicago: U of Chicago P, 1975) 24.

25. Abrams, *Natural Supernaturalism* 35.

26. Yates, *Theatre of the World* 128, 134.

27. Jules Michelet, *History of the French Revolution,* trans. Charles Cocks (Chicago: U of Chicago P, 1967) 3. White, in *Metahistory,* also suggests that Michelet turns history into

a morality play, arguing that he emplotted history as a drama of "a spiritual power fighting to free itself from the forces of darkness" (p. 152).

28. R. G. Collingwood, *The Idea of History* (London: Oxford UP, 1970) 39.

29. Wilhelm Dilthey, *Pattern and Meaning in History*, ed. H. P. Rickman (New York: Harper and Row, 1961) 67–68.

30. David Bromwich, *William Hazlitt: The Mind of a Critic* (New York: Oxford UP, 1983) 373.

31. Avrom Fleishman, *The English Historical Novel* (Baltimore: John Hopkins UP, 1971) xii.

32. Robert Langbaum, *The Poetry of Experience* (New York: Random House, 1957) 25.

33. Jean-Paul Sartre, *Search for a Method*, trans. Hazel E. Barnes (New York: Alfred A. Knopf, 1963) 45–46; italics his.

34. Ibid., 46; italics Sartre's.

35. Georges Lefebvre, *The French Revolution, from Its Origin to 1793*, trans. Elizabeth Moss Evanson (1957; New York: Columbia UP, 1962) 149. This is the first of two volumes. Vol. 2 was published in 1964.

36. Fredric Jameson, *Marxism and Form* (Princeton: Princeton UP, 1971) 223.

37. Paul Ricoeur, *Time and Narrative*, trans. Kathleen McLaughlin and David Pellauer, 3 vols. (Chicago: U of Chicago P, 1984) 1: 82. Vol. 2 published 1986; vol. 3 forthcoming.

38. Hayden White, *The Tropics of Discourse* (Baltimore: Johns Hopkins UP, 1978) 82.

39. Ricoeur, *Time and Narrative* 1: 3.

40. Claude Lévi-Strauss, *The Savage Mind* (Chicago: U of Chicago P, 1966) 257.

41. This difference between scientific fact and historical fact is stressed by W. Wolfgang Holdheim in *The Hermeneutic Mode: Essays on Time in Literature and Literary Theory* (Ithaca, N.Y.: Cornell UP, 1984) 216–17.

42. Burke, *Attitudes toward History* 20–21.

43. John C. Greene, *Science, Ideology, and World View: Essays in the History of Evolutionary Ideas* (Berkeley: U of California P, 1981) 49.

44. Collingwood, *The Idea of History* 94–95.

45. Quoted in Greene, *Science, Ideology, and World View* 107.

46. R. R. Palmer, *The Age of the Democratic Revolution*, 2 vols. (Princeton: Princeton UP, 1959, 1964) 1: 308, 309.

47. Dominick La Capra, *Rethinking Intellectual History* (Ithaca, N.Y.: Cornell UP, 1983) 18.

48. Ricoeur, *Time and Narrative* 1: 99.

49. See Roland Barthes, "Historical Discourse," in Michael

Lane, ed., *Introduction to Structuralism* (New York: Basic Books, 1970) 153.

50. White, *The Tropics of Discourse* 112.
51. Barthes, "Historical Discourse" 149.
52. White, *The Tropics of Discourse* 112.
53. Ricoeur, *Time and Narrative* 1: 76, 78.
54. Gerald Graff, *Literature against Itself* (Chicago: U of Chicago P. 1979) 86–87; E. D. Hirsch, Jr., *The Aims of Interpretation* (Chicago: U of Chicago P, 1976) 8.
55. J.R.R. Tolkien, "On Fairy Stories," *Tree and Leaf* (Boston: Houghton Mifflin, 1965) 30; Tolkien, *The Lord of the Rings* (New York: Ballantine Books, 1965) xi (the "authorized edition").
56. Sir Walter Scott, *Essays on Chivalry, Romance, and the Drama* (Freeport, N.Y.: Books for Libraries Press, 1972) 134–35.
57. Frye, *Anatomy of Criticism* 105.
58. Arnold Toynbee, *A Study of History*, ed. and abr. D. C. Somervell, 2 vols. (New York: Oxford UP, 1946, 1957) 1: 44.
59. Ibid., 60.
60. E. H. Carr, *What Is History?* (New York: Alfred A. Knopf, 1961) 59, 65.
61. Jameson, *Marxism and Form* 221; Sartre, *Search for a Method* 44.
62. Carr, *What is History?* 165–66.
63. Toynbee, *A Study of History* 1: 60.
64. Frye, *Anatomy of Criticism* 148.
65. Ibid., 63.
66. Paulson, *Representations of Revolution* 8.
67. Alexis de Tocqueville, *The Old Regime and the French Revolution*, trans. Stuart Gilbert (1856; Garden City, N.Y.: Doubleday, 1955) 19.
68. Frye, *Anatomy of Criticism* 148.
69. Graff, *Literature against Itself* 182.
70. Paul de Man, "Literary History and Literary Modernity," *Blindness and Insight* (Minneapolis: U of Minnesota P, 1983), especially pp. 150, 151, 161, 163, 165.
71. Herbert Marcuse, *Eros and Civilization: A Philosophical Inquiry into Freud* (Boston: Beacon P, 1955) 213; and Walter Benjamin, *Illuminations*, ed. Hannah Arendt (New York: Harcourt, Brace and World, 1955) 263. Their view of time as an instrument of tyranny is discussed in Martin Jay, *Marxism and Totality: The Adventures of a Concept from Lukács to Habermas* (Berkeley: U of California P, 1984) 234.
72. Marcuse, *Eros and Civization* 213; quoted in Jay, *Marxism and Totality* 234.

73. Tocqueville, *The Old Regime and the French Revolution* 146.
74. Ibid., ix.
75. Palmer, *The Age of the Democratic Revolution* 1: 441.

Chapter Two

1. Charles Daubuz, *A Perpetual Commentary on the Revelation of St. John* (London, 1720) 56.
2. Blake's reaction against a world view predicated on mathematically measured time and space has been discussed at length, most notably perhaps by Northrop Frye in *Fearful Symmetry* (Princeton: Princeton UP, 1947); see especially p. 46. W.J.T. Mitchell, in *Blake's Composite Art* (Princeton: Princeton UP, 1978) 34, argues that Blake's poetry functions to invalidate objective time and that his painting functions to invalidate objective space. Leslie Tannenbaum, in *Biblical Tradition in Blake's Early Prophecies* (Princeton: Princeton UP, 1982) 87, suggests that Blake distinguishes between history as *kairos* (discrete moments, each having intemporal, even divine, significance) and history as *chronos* (clock time).
3. C. A. Patrides, *The Grand Design of God* (London: Routledge and Kegan Paul, 1972) 28.
4. Thomas Paine, *Common Sense* (Newport, 1776), unpaginated. Richard Price, *Observations on the Importance of the American Revolution, The Works of Richard Price*, 10 vols. (London, 1816) 8: 3.
5. Patrides, *The Grand Design of God* 49–50.
6. J. Bicheno, *The Signs of the Times, or the Overthrow of Papal Tyranny in France* (London, 1793) 93; Bicheno, *The Probable Progress and Issue of the Commotions Which Have Agitated Europe since the French Revolution* (London, 1797) 1.
7. Alexander Pirie, *The French Revolution Exhibited in the Light of the Sacred Oracles* (Perth, 1795) 49, 63. Italics his.
8. Gerrard Winstanley, *The New Law of Righteousness, The Works of Gerrard Winstanley*, ed. George H. Sabine (New York: Russell and Russell, 1965) 176–77.
9. Ibid., 230.
10. See Tannenbaum, *Biblical Tradition*, 88.
11. Ibid., 137; italics Lee's.
12. David V. Erdman, *Blake: Prophet against Empire* (Garden City, N.Y.: Doubleday, 1969) 58.
13. Bicheno, *Commotions Which Have Agitated Europe*, 55.
14. Erdman, *Blake: Prophet against Empire* 58–59.

15. Joshua Barnes, *The History of that Most Victorious Monarch Edward III* (Cambridge, England, 1688) 341.

16. White, *The Tropics of Discourse* 5, 91.

17. Joseph A. Wittreich, *Visionary Poetics* (San Marino, Calif.: Huntington Library, 1979) 14. For a discussion of how the reader participates in building analogies to the thing itself, see Wolfgang Iser, *The Act of Reading* (Baltimore: Johns Hopkins UP, 1978) 64.

18. See Mitchell, *Blake's Composite Art* 31–35, for a brilliant, if generalized, account of this process in the Prophetic Books.

19. Tannenbaum connects the portraits with this kinship. See *Biblical Tradition* 47.

20. George Quasha, "Orc as a Fiery Paradigm of Poetic Torsion," in David V. Erdman and John E. Grant, eds., *Blake's Visionary Forms Dramatic* (Princeton: Princeton UP, 1970) 264, 275. Nelson Hilton elaborates on this function of words in Blake's works; see *Literal Imagination: Blake's Vision of Words* (Berkeley: U of California P, 1983), especially p. 2.

21. Lawrence Lipking treats this claim as the informing theme of *The Marriage of Heaven and Hell* in his fine book, *The Life of the Poet* (Chicago: U of Chicago P, 1981) 34–47.

22. For Erdman's interpretation, which is, so far as I know, universally accepted, see *Blake: Prophet against Empire* 24–25.

23. Tannenbaum, *Biblical Tradition* 50. Tannenbaum cites Ezekiel 7:1–15 and Isaiah 21:12.

24. Hilton, *Literal Imagination* 28.

25. David V. Erdman, "*America*: New Expanses," in Erdman and Grant, *Blake's Visionary Forms Dramatic* 94–95; Tannenbaum, *Biblical Tradition* 47.

26. See especially Mitchell, *Blake's Composite Art*, p. 31.

27. Kermode, *The Sense of an Ending* 67–89.

28. I have followed Erdman's chronology, which seems to me well documented and persuasive. See "*America*: New Expanses" 94. See also M. H. Abrams's observation, in *Natural Supernaturalism* 257, that all of Blake's epics represent the biblical plot of Creation, Fall, human history, and (by anticipation) redemption in a restored Eden.

29. Quoted in Wittreich, *Visionary Poetics* 24.

30. Paul Henri Mallet, *Northern Antiquities*, trans. Bishop Percy, 2 vols. (London, 1752) 1: 113–14.

31. Jacob Bryant, *A New System, or an Analysis of Ancient Mythology*, 3 vols. (London, 1774) 1: 326–27.

32. Lipking, *The Life of the Poet* 43.

33. John Milton, *The History of Britain*, ed. George Philip Krapp. This is vol. 10 of *The Works of John Milton*, ed.

Frank Allen Patterson, 18 vols. (New York: Columbia UP, 1931–1938) 10 (1932): 103. Italics Milton's.

34. Michael Tolley, "*Europe*: 'to those ychain'd in sleep'," in Erdman and Grant, *Blake's Visionary Forms Dramatic* 116.

35. Arthur O Lovejoy, *The Great Chain of Being* (Cambridge, Mass.: Harvard UP, 1936) 203–4.

36. Tannenbaum, *Biblical Tradition* 167.

37. Erdman, *Blake: Prophet against Empire* 211, 222–23.

38. M. H. Abrams, *The Mirror and the Lamp: Romantic Theory and Critical Tradition* (New York: Oxford UP, 1953) 264.

39. Christopher Hill, *The World Turned Upside Down* (New York: Viking, 1972) 70, 237.

40. Erdman, *Blake: Prophet against Empire* 224.

Chapter Three

1. Hazlitt's remark to Medwin (on learning in 1825 of Scott's projected *Life*) that "I too will write a Life of Napoleon, though it is yet too early," is quoted by Howe in a note to Hazlitt's *Complete Works* (13: 356).

2. Bromwich, *Hazlitt: The Mind of a Critic*, 288.

3. Ibid.

4. That Burke is the source of Hazlitt's label "Iliad of woes" for the First Coalition's struggle, first against the republic, then against Napleon, is documented several times by Howe in his notes to Hazlitt's *Complete Works.*

5. For the tale of Napoleon's birth, apparently as he himself told it, see Count de Las Cases, *Memories of the Life, Exile, and Conversations of Napoleon*, 4 vols. (1823; New York, 1894) 1: 73.

6. Bromwich, *Hazlitt: The Mind of a Critic* 291.

7. Ibid., 294.

8. Ibid.

9. Mark A. Weinstein, "Sir Walter Scott's French Revolution: The British Conservative View," *Scottish Literary Journal* 7 (1980): 34.

10. Bromwich, *Hazlitt: The Mind of a Critic* 294; italics Bromwich's.

11. Ibid., 303.

12. Pieter Geyl, *Napoleon: For and Against,* trans. Olive Renier (New Haven: Yale UP, 1949) 9.

13. See *Essays on Chivalry, Romance, and the Drama* 11 for Scott's strictures on chivalry.

14. Paulson, *Representations of Revolution* 8, 10.

15. Though Hazlitt avoids identifying the target of his invec-

tive here, it is easily recognizable as Scott; see Howe's note about this passage in Hazlitt's *Complete Works* (15: 380).

16. Bromwich, *Hazlitt: The Mind of a Critic* 278.

17. The allusion is identified by Howe in a note to Hazlitt's *Complete Works* (14: 367).

18. Bromwich, *Hazlitt: The Mind of a Critic* 281.

19. Robert C. Gordon, "Scott among the Partisans: A Significant Bias in His 'Life of Napleon Buonaparte,' " *Scott Bicentenary Essays*, ed. Allen Bell (Edinburgh: Scottish Academic P, 1973) 120, 123.

20. See the translation of Segur by J. David Townsend, published as *Napoleon's Russian Campaign* (Boston: Houghton Mifflin, 1958) 52.

21. *Memoirs of the Duke of Rovigo*, 4 vols. (London, 1828) 1: i, 308.

22. See Earl Stanhope, *Life of the Right Honorable William Pitt* (London, 1862) 4: 369; Lord Rosebery, *Pitt* (1891; London: Macmillan, 1908) 256.

23. Howe identifies the phrase Hazlitt has quoted (13: 360).

24. For this idea, I am indebted to my colleague Professor Phillips Salman.

25. Howe (13: 359) traces Hazlitt's echo of Milton to *Paradise Lost* 4: 17. Might Hazlitt also have had in mind the American revolutionary flag, which bore the motto "Don't tread on me"?

26. Howe (15: 379) cites *Paradise Lost* 4: 799–800.

27. Harry E. Shaw, *The Forms of Historical Fiction* (Ithaca, N.Y.: Cornell UP, 1983) 190.

28. See Christopher Dawson on *La Chanson de Roland* in *Religion and the Rise of Western Culture* (Garden City, N.Y.: Doubleday, 1958) 145.

Chapter Four

1. Michelet, *History of the French Revolution* 3; Albert Mathiez, *The French Revolution*, trans. Catherine Alison Phillips (New York: Russell and Russell, 1962) 14. That Carlyle understood the French Revolution as an event as much of the mind as of the political and social world is one of the main theses of John Rosenberg's important book, *Carlyle and the Burden of History* (Cambridge, Mass.: Harvard UP, 1985).

2. *Two Notebooks of Thomas Carlyle*, ed. Charles Eliot Norton (New York: Grolier Club, 1938) 124; italics Carlyle's. John Rosenberg, in *Carlyle and the Burden of History* 44, identifies Carlyle's main concern with the linearity of nar-

rative as "how to depict beginnings and endings, or even coherent middles, when discrete beginnings and endings are chimeras of the mind and continuity itself is inherently resistant to verbal representation." He is right, but he seems seriously to understate what, for Carlyle, is the principal problem: how to depict not continuity but simultaneity. Rosenberg's argument, particularly as it concerns Carlyle's method of ending, will be examined more fully in the Conclusion.

3. I have borrowed the term "synoptically" from H. M. Leicester, Jr., who borrowed it from Charles Frederick Harold. Leicester's essay, "The Dialectic of Romantic Historiography: Prospect and Retrospect in 'The French Revolution,'" *Victorian Studies* 15 (1971): 5–17, is one of the best studies to date of Carlyle's narrative technique in *The French Revolution*.

4. Leicester, "The Dialectic of Romantic Historiography" 6–7.

5. Michael Goldberg, in *Carlyle and Dickens* (Athens: U of Georgia P, 1972) 121, links Fortunatus' hat to Carlyle's manipulation of point of view in *The French Revolution*. But Goldberg is concerned with describing the influence of Carlyle on Dickens, not with examining Carlyle's narrative strategy.

6. Elliot L. Gilbert, " 'A Wondrous Contiguity': Anachronism in Carlyle's Prophecy and Art," *PMLA* 87 (1972): 434.

7. Gilbert, "A Wondrous Contiguity" 436. See also John Rosenberg, *Carlyle and the Burden of History* 30.

8. John Rosenberg (*Carlyle and the Burden of History*, 58) suggests that the effect of present tense narration is to "[remove] the distancing frame from the historical picture and [thrust] both narrator and reader into the field of action."

9. Collingwood, *The Idea of History* 39. Collingwood's description of his experience in writing about Nelson is quoted in Thomas R. Whitaker, *Swan and Shadow: Yeats's Dialogue with History* (Chapel Hill: U of North Carolina P, 1959) 204.

10. There is no evidence that Carlyle ever read Blake. But as Albert J. LaValley points out in *Carlyle and the Idea of the Modern* (New Haven: Yale UP, 1968) 164–70, the parallels between them are striking.

11. *The Collected Letters of Thomas and Jane Welsh Carlyle*, ed. Charles R. Sanders and Kenneth J. Fielding, 9 vols. (Durham, N.C.: Duke UP, 1970–1981) 4: 446; italics Carlyle's.

12. Abrams, *Natural Supernaturalism* 330–32, 301.

13. Norton, *Two Notebooks* 187–88; italics Carlyle's.

14. *The Literary Notes of Thomas Hardy*, ed. Lennart A. Björk, 2 vols. (Göteborg, Sweden: Acto Universitatis Gothenbur-gensis, 1974) 1 (text): 167.

15. Leicester, "The Dialectic of Romantic Historiography" 15.

16. Lévi-Strauss, *The Savage Mind* 257.

17. John Rosenberg (*Carlyle and the Burden of History* 110) views the ending of Carlyle's narrative differently. To Rosenberg, the essential point is that the French Revolution is, "by nature," without end, as universal history is without end; like Homer's Epos, it "merely ceases."

18. Carlyle, *Collected Letters* 6: 302.

19. For Mill's objection to Carlyle's evaluation, see his letter to Carlyle of 2 February 1833 in *The Early Letters of John Stuart Mill*, ed. Francis E. Mineka, 2 vols. (Toronto: U of Toronto P, 1963) 1: 139. This volume is volume 12 of the *Collected Works of John Stuart Mill*, ed. J. M. Robson *et al*, 25 vols. (Toronto: U of Toronto P, 1963–86). Carlyle's reassessment appears in an article on Necker written for the *Edinburgh Encyclopedia*. See *Works* 30: 94.

20. Carlyle, *Collected Letters* 6: 302–3.

21. Geoffrey Hartman, *The Fate of Reading* (Chicago: U of Chicago P, 1975) 119; Edward Said, *Beginnings* (New York: Basic Books, 1975) 29.

22. Janet Ray Edwards alludes to, but never really examines, the importance of processions in *The French Revolution*; see "Carlyle and the Fictions of Belief: *Sartor Resartus* to *Past and Present*," in John Clubbe, ed., *Carlyle and His Contemporaries* (Durham, N.C.: Duke UP, 1976) 103. John D. Rosenberg (*Carlyle and the Burden of History* 72) asserts, I think rightly, that *The French Revolution* is plotted around a series of royal processions.

23. Charles Frederick Harrold, "Carlyle's General Method in *The French Revolution*," *PMLA* 43 (1928): 1152.

24. Madame Campan, *Memoirs of the Private Life of Marie Antoinette, Queen of France and Navarre*, 2 vols. (London, 1823) 1: 47.

25. Clubbe, *Carlyle and His Contemporaries* 98.

26. Campan, *Private Life of Marie Antoinette* 1: 409.

27. P. L. Roederer, "Chronique de Cinquante Jours, du 20 juin au 10 août 1792," M. de Lescure, ed., *Bibliothèque des Mémoires relatif à l'Histoire de France pendant de 18e siècle* (Paris, n.d.) 84–88. Roederer's account of his heroics outside the Salle de Manège reads: "je m'avancai, et montant sur la quatrième marche de l'escalier, je dis: 'Citoyens, je vous demande du silence au nom de la loi.' J'obtins du si-

lence. . . . L'opposition générale parut céder. Mais l'homme à la longue perche la brandissait en criant toujours: '*à bas! à bas!*' Je montai sur la terrasse, la lui arrachai des mains et la jetai dans le jardin. L'étonnement l'empêcha de crier davantage" (pp. 87–88).

28. John D. Rosenberg, in *Carlyle and the Burden of History* 41, observes that what men believe and fear Carlyle took to be as integral to history as what they see and do. He cites Carlyle's decision not to remove the fabricated details of the *Vengeur*'s sinking from later editions of *The French Revolution* as evidence of this conviction.

29. Roederer, "Chronique de Cinquante Jours" 83; italics his.

30. "M. Isnard, s'étant fait élever sur les épaules de deux volontaires, parla le premier. . . . M. Vergniaud prit à son tour la parolle, et parvint à faire écouter un discours assez long dans lequel il rappelait le respect dû aux autorités, et tâchait de faire entendre que ce n'était pas par la violence qu'il était possible d'obtenir ce qu'on demandait" (pp. 51–52).

31. Roederer, "Chronique de Cinquante Jours" 48–49. Campan, in *Private Life of Marie Antoinette* 2: 209, corroborates Roederer's version.

32. "Il [Louis] passe de sa chambre dans son cabinet, de là à la chambre du lit et à l'Oeil-de Boeuf, accompagné de madame Elisabeth, de trois ministres: MM. Beaulieu, de Lajeard et Terrier. M. le maréchal de Mouchy, MM. d'Hervilliers et de Canolle, M. Guinguerlot, lieutenant-colonel de la gendarmerie à pied, et M. de Vainfrais, autre officier de gendarmerie, se réuinissent auprès du roi" (p. 47).

33. M. Mercier, *New Picture of Paris*, 2 vols. (Dublin, 1800) 2: 173–74.

34. Ibid., 173.

35. John Rosenberg, *Carlyle and the Burden of History* 100, emphasizes dream as one of Carlyle's chief metaphors for the Revolution.

36. Morse Peckham, *Beyond the Tragic Vision* (New York: George Brazillier, 1962) 15.

37. Yeats, *Letters* 922.

38. White, *The Tropics of Discourse* 64.

39. LaValley, *Carlyle and the Idea of the Modern* 7, 122–23. See also Philip Rosenberg, *The Seventh Hero: Thomas Carlyle and the Theory of Radical Activism* (Cambridge, Mass.: Harvard UP, 1974) 75.

40. Philip Rosenberg, *The Seventh Hero* 75.

41. John Rosenberg has largely anticipated me in this argu-

ment, though he frames it rather differently. See *Carlyle and the Burden of History* 97.

42. Angus Fletcher, *Allegory: The Theory of a Symbolic Mode* (Ithaca, N.Y.: Cornell UP, 1964) 35, 39.

43. Philip Rosenberg, *The Seventh Hero* 168.

Chapter Five

1. Walter Bagehot, *The Literary Essays*, ed. Norman St. John–Stevas, 2 vols. (Cambridge, Mass.: Harvard UP, 1965) 2: 87.

2. Goldberg, *Carlyle and Dickens* 105, calls attention to the parallels in the narrative strategies of *A Tale of Two Cities* and *The French Revolution*.

3. Bagehot, *The Literary Essays* 2: 87.

4. Taylor Stoehr, *Dickens: The Dreamer's Stance* (Ithaca, N.Y.: Cornell UP, 1965) 11.

5. Jonathan Arac, *Commissioned Spirits: The Shaping of Social Motion in Dickens, Carlyle, Melville, and Hawthorne* (New Brunswick, N.J.: Rutgers UP, 1979) 2.

6. Rudolf Bultmann, *History and Eschatology* (Edinburgh: At the UP, 1957) 119.

7. Robert Alter, "The Demons of History in Dickens' *Tale*," *Novel* 2 (1969): 137.

8. T. A. Jackson, in *Charles Dickens: The Progress of a Radical* (1937; New York: Haskell House, 1971), argues that what he calls the conventional view of *A Tale of Two Cities*—which identifies Carton's sacrifice as its primary thematic objective—is false criticism, because it relegates the Revolution and the subplot of Madame Defarge's revenge to irrelevancy. But the conventional view, though it treats Madame Defarge's revenge as a subplot, does not relegate the Revolution to irrelevancy. And if Jackson wants to read Carton's sacrifice as subordinate to Madame Defarge's revenge, he must restructure the novel. His argument is ideological: conventional readers are wrong critically because they are wrong politically; they distort not the emphases established by the rhetorical strategy of the novel but those established by Jackson's own ideologically determined priorities. Orwell dismisses Jackson's argument in his essay, "Charles Dickens," reprinted in George Orwell, *An Age Like This*, vol. 1 of *The Collected Essays, Journalism and Letters of George Orwell*, ed. Sonia Orwell and Ian Angus, 4 vols. (New York: Harcourt, Brace and World, 1968) 416.

9. Jameson, *The Political Unconscious* 102; italics his.

10. Kermode, *The Sense of An Ending* 72.

11. See Woodcock's introduction to the Penguin *A Tale of Two Cities* 11.

12. Ibid., 23.

13. John Kucich notes the contrast between Lucie's golden thread and Madame Defarge's knitting in "The Purity of Violence: *A Tale of Two Cities*," *Dickens Studies Annual*, ed. Michael Timko, Fred Kaplan, and Edward Guilano (New York: AMS Press, 1980) 8: 127.

14. For Alter's interpretation, see "The Demons of History" 139.

15. The list of studies devoted to Carlyle's influence on Dickens, and not only in *A Tale of Two Cities*, has grown over the years. The fullest treatments are Goldberg, *Carlyle and Dickens*; and William Oddie, *Dickens and Carlyle: The Question of Influence* (London: Centenary P), both published in 1972. But several articles have also been devoted partly or entirely to this subject, most notably perhaps John P. McWilliams, Jr., "Progress without Politics: *A Tale of Two Cities*," *Clio* 7 (1977): 19–31; and David D. Marcus, "The Carlylean Vision of *A Tale of Two Cities*," *Studies in the Novel* 8 (1976): 56–58. There are chapters or sections of chapters on the subject in some of the more general studies of Dickens as well.

16. Stoehr, *The Dreamer's Stance* 66, 69, 203.

17. Hans Meyeroff, *Time in Literature* (Berkeley: U of California P, 1968) 4, 15.

18. See Alter, "The Demons of History" 139.

19. Quoted in John Carey, *The Violent Effigy: A Study of Dickens' Imagination* (London: Faber and Faber, 1973) 20–21.

20. Philip Collins, in *Dickens and Crime* (London: Macmillan, 1964) 132–33, identifies Dickens' visit to Cherry Hill as a source for his depiction of Dr. Manette's imprisonment. The quotation is from Dickens' *American Notes* and is cited by Collins (p. 132).

21. *The Selected Letters of Charles Dickens*, ed. F. W. Dupee (New York: Farrar, Straus, and Cudahy, 1960) 256–57.

22. Oddie, *Dickens and Carlyle* 35, calls attention to the parallels in imagery, as well as incident, between the accounts of the storming of the Bastille in *A Tale* and *The French Revolution*.

23. Laing, *The Divided Self* 79.

24. Arac, *Commissioned Spirits* 124. Arac accepts P. P. Howe's identification of Burke as the source of the phrases Hazlitt quotes.

25. Miyoshi, *The Divided Self* 5.

26. Northrop Frye, *The Secular Scripture* (Cambridge, Mass.: Harvard UP, 1976) 53.

27. Lawrence Frank, "Dickens' *A Tale of Two Cities*: The Poetics of Impasse," *American Imago* 36 (1979): 229.

28. Edwin Eigner, in *The Metaphysical Novel in England and America* (Berkeley: U of California P, 1978) 136, calls Charles a guilty hero. Most critics of *A Tale of Two Cities* point out, moreover, that Charles's family name in English means "Everyman." But they are curiously incurious about why.

29. Stoehr, *The Dreamer's Stance* 198.

30. Frye, *The Secular Scripture* 143.

31. John Kucich, *Excess and Restraint in the Novels of Charles Dickens* (Athens: U of Georgia P, 1981) 117. The interpretation of "Lorry" as a pun especially is a compelling idea. But the usages listed by the OED as current in the mid-nineteenth century indicate a vehicle sufficiently different from a tumbrel to raise questions.

32. Georg Lukács, *The Historical Novel*, trans. Hannah and Stanley Mitchell (Boston: Beacon Press, 1963) 243. This seems to me a very curious complaint for Lukács to make, since it describes, in greater or lesser degree, the general strategy of historical novels. Frank examines Lukács's argument critically in "Dickens' *A Tale of Two Cities*" 215–16.

33. Albert D. Hutter, "Nation and Generation in *A Tale of Two Cities*," *PMLA* 93 (1978): 448. Hutter calls these strains a revolt against authority, which strikes me as too strong.

34. Hutter, "Nation and Generation" 450.

35. Ibid., 454.

36. *The Speeches of Charles Dickens*, ed. K. J. Fielding (Oxford: Clarendon P) 197.

37. Greene, *Science, Ideology, and World View*, 60

38. Herbert Spencer, *First Principles* (1862; New York, 1880) 297.

39. Spencer, *First Principles* 450.

40. Kucich, *Excess and Restraint* 117.

Chapter Six

1. Thomas Hardy, *Notes on "The Dynasts" in Four Letters to Edward Clodd* (Edinburgh: Dunedin P, 1929) 13.

2. See Walter de la Mare, *Private View* (London: Faber and Faber, 1953) 24; Emma Clifford, "The Impressionistic View of History in *The Dynasts*," *Modern Language Quarterly*

22 (1961): 22; J. Hillis Miller, *Thomas Hardy: Distance and Desire* (Cambridge, Mass.: Harvard UP, 1970) 8; and Susan Dean, *Hardy's Poetic Vision in "The Dynasts"* (Princeton: Princeton UP, 1977) 31 on the Spirits as mental constructs. See also Hardy, *Notes on "The Dynasts"* 8. This letter is quoted in Florence Emily Hardy, *The Life of Thomas Hardy* 301–2, where it is misdated 1905.

3. Edward Clodd, *Pioneers of Evolution from Thales to Huxley* (New York, 1897) 232–33.

4. I am following here the analysis of mechanistic reasoning in Stephen C. Pepper, *World Hypotheses* (Berkeley: U of California P, 1942) 196.

5. Henry Thomas Buckle, general introduction to *History of Civilization in England*, in *Varieties of History* 124–25.

6. Quoted in Clifford, "The Impressionistic View of History" 26.

7. Quotations from *The Dynasts* are cited by part, act, scene, and page numbers.

8. Hardy expressed his praise of Spencer's *First Principles* in a letter dated July 17, 1893. See *The Collected Letters of Thomas Hardy*, ed. Richard Little Purdy and Michael Millgate, 5 vols. thus far (Oxford: Clarendon P, 1978–85) 2: 24.

9. Charles Darwin, *The Origin of Species by Means of Natural Selection* (1859; New York: Random House, n.d.) 64.

10. For White's suggestion that *The Origin of the Species* should be read as an artistic construct, see *The Tropics of Discourse* 133. For Hyman's argument, see *The Tangled Bank* (New York: Atheneum, 1962) 26–29.

11. Langbaum examines the parallel between romantic epistemology and scientific method in *The Poetry of Experience* 22.

12. White, *The Tropics of Discourse* 131; italics his.

13. Darwin, *The Origin of Species* 135, 184.

14. White, *The Tropics of Discourse* 131.

15. Langbaum, *The Poetry of Experience* 22.

16. Dean, *Hardy's Poetic Vision* 68.

17. White, *The Tropics of Discourse* 98. See also White, *Metahistory* x.

18. White, *Metahistory* 27, 178; White, *The Tropics of Discourse* 59, 110.

19. Frye, *Anatomy of Criticism* 15; Frye, *Fables of Identity* 31.

20. White, *The Tropics of Discourse* 53.

21. John Wain, introduction to *The Dynasts* ix. Hardy seems to have anticipated the parallel Siegfried Kracauer perceives between reality viewed through the lens of a camera and reality viewed through the mind's eye. Each reality is part given, part composed; each depends for its coherence on the imposition of a frame that delimits the subject. See

History: The Last Things Before the Last (New York: Oxford UP, 1969) 59. Hardy also anticipates Mircea Eliade's perception, in *Myths, Dreams, and Mysteries*, trans. Philip Mariet (New York: Harper and Row, 1960) 34, that cinematic technique condenses time. Dean (*Hardy's Poetic Vision* 19) recognizes, though only in a rudimentary way, Hardy's manipulation of time, labeling his treatment of history "thin" and recommending that it be approached as a series of visual impressions of past moments. She essentially repeats the thesis of Emma Clifford who, in "The Impressionistic View of History," characterizes Hardy's treatment of history as vague—a judgment that strikes me as seriously flawed.

22. Hardy, *Literary Notes* 1: 152.

23. Miller, *Thomas Hardy: Distance and Desire* viii.

24. Hardy, *Literary Notes* 1: 167.

25. Pitt's speech is quoted almost as Hardy reproduces it in Stanhope, *William Pitt* 4: 346; and in Rosebery, *Pitt* 255, both of which Hardy read. His extensive research for *The Dynasts* is traced in Walter F. Wright, *The Shaping of "The Dynasts"* (Lincoln: U of Nebraska P, 1967).

26. Scott, *Life of Napoleon Buonaparte* 2: 223; W.F.P. Napier, *History of the War in the Peninsula and in the South of France*, 5 vols. (1862; New York: AMS Press, 1970) 1: 225–26.

27. Napier, *History of the War* 1: 337–38; and James C. Moore, *The Life of Lieutenant-General Sir John Moore, K.B.*, 3 vols. (London, 1834) 2: 221–30.

28. Fussell, *The Great War and Modern Memory* ix.

29. William Beatty, *Authentic Narrative of the Death of Lord Nelson* (London, 1807) 48.

30. Ibid., 36.

31. Napier, *History of the War* 1: 339.

32. Quoted in Wright, *The Shaping of "The Dynasts"* 80.

33. Stanhope, *William Pitt* 4: 369; Rosebery, *Pitt* 256; W. M. Sloane, *Life of Napoleon Bonaparte*, 4 vols. (London: Macmillan, 1910) 2: 254.

34. See Sloane, *Life of Napoleon Bonaparte* 4: 75.

35. Pierre Lanfrey, *History of Napoleon the First*, 4 vols. (London, 1894) 4: 94.

36. John Holland Rose, *The Life of Napoleon I*, 2 vols. (New York: Macmillan, 1902) 2: 243.

37. Tolkien, *Tree and Leaf* 68.

38. Clifford cites this passage (see "The Impressionistic View of History" 31) but does no justice to its implications.

39. Henry Houssaye, *1815: Waterloo*, trans. Arthur Emile Mann (London: Adam and Charles Black, 1900) 83.

40. Jean R. Brooks, *Thomas Hardy: The Poetic Structure* (Ithaca, N.Y.: Cornell UP, 1971) 278.
41. Keegan, *The Face of Battle* 178–79.

Conclusion

1. John Emerich Edward Dalberg-Acton, *Lectures on the French Revolution*, ed. John Neville Figgis and Reginald Vere Laurence (New York: Noonday, 1959) 74; Georges Lefebvre, *The French Revolution, from 1793 to 1799*, trans. John Hall Stewart and James Friguglietti (New York: Columbia UP, 1964) 256. This is volume 2 of Lefebvre's history of the Revolution.
2. Lefebvre, *The French Revolution* 2: 145.
3. Barbara Herrnstein Smith, *Poetic Closure: A Study of How Poems End* (Chicago: U of Chicago P, 1968) 120.
4. Ibid.
5. Robert M. Adams, *Strains of Discord: Studies in Literary Openness* (Ithaca, N.Y.: Cornell UP, 1958) 15; J. Hillis Miller, "Narrative Endings," *Nineteenth-Century Fiction* 33 (1978): 6.
6. Holdheim, *The Hermeneutic Mode* 202.
7. John D. Rosenberg, *Carlyle and the Burden of History* 44.
8. Ibid., 110.
9. Tocqueville states unequivocally that "the French monarchy, after being swept away by the tidal wave of Revolution, was restored in 1800." *The Old Regime and the French Revolution* 60.
10. Ibid., xi.
11. Lefebvre, *The French Revolution* 2: 317.
12. Lefebvre, *The French Revolution* 1: 79.
13. Jay, *Marxism and Totality* 316.
14. Abrams, *Natural Supernaturalism* 313–14.
15. Toynbee, *A Study of History* 1: 187, 253.
16. Benjamin, *Illuminations* 257.
17. See Michel Foucault, *Language, Counter-Memory, Practice: Selected Essays and Interviews*, ed. Donald F. Bouchard, trans. B. and Sherry Simon (Ithaca, N.Y.: Cornell UP, 1977) 85; and Foucault, *The Archaeology of Knowledge* 9. See also Jay, *Marxism and Totality* 521.
18. Benjamin, *Illuminations* 264.
19. Foucault, *The Archaeology of Knowledge* 110; Benjamin, *Illuminations* 264.
20. Derrida, *Of Grammatology* 36. Ryan, *Marxism and Deconstruction* 20, also recognizes that the deconstructive approach to narrative leads logically to endless regress. But I

find little evidence that he appreciates the problem this view creates for understanding, or even taking seriously, anything written, including his own book.

21. Graff, *Literature against Itself* 11.
22. Jameson, *Marxism and Form* 38.
23. Derrida, *Of Grammatology* 35.
24. Burke, *Attitudes toward History* 159.
25. Sartre, *Search for a Method* 138.
26. La Capra, *Rethinking Intellectual History* 281.
27. Sartre, *Search for a Method* 138, italics his.
28. Owen Barfield, *Romanticism Comes of Age* (Middletown, Conn.: Wesleyan UP, 1967) 39-40.

Index

Barton Friedman demonstrates the ways in which English men of letters in the nineteenth century attempted to grasp the dynamics of history and to fashion order, however fragile, out of its apparent chaos. The authors he discusses—Blake, Scott, Hazlitt, Carlyle, Dickens, and Hardy—found in the French Revolution an event more compelling as a paradigm of history than their own "Glorious Revolution." To them the French Revolution seemed universally significant—a microcosm, in short. For these writers maintaining the distinction between "history" and "fiction" was less important than making sense of epochal historical events in symbolic terms. Their works on the French Revolution and the Napoleonic Wars occupy the boundary between history and fiction, and *Fabricating History* advances the current lively discussion of that boundary.

At the same time, this work explores questions about narrative strategies, as they are shaped by, or shape, events. Narratives incorporate the ideological and metaphysical preconceptions that the authors bring with them to their writing. "This is not to argue," Professor Friedman says, "that historical narratives are only about the mind manufacturing